Project AIR FORCE Modeling Capabilities for Support of Combat Operations in Denied Environments

Brent Thomas, Mahyar A. Amouzegar, Rachel Costello, Robert A. Guffey,
Andrew Karode, Christopher Lynch, Kristin F. Lynch, Ken Munson, Chad J. R. Ohlandt,
Daniel M. Romano, Ricardo Sanchez, Robert S. Tripp, Joseph V. Vesely

RAND Project AIR FORCE

Prepared for the United States Air Force
Approved for public release; distribution unlimited

For more information on this publication, visit www.rand.org/t/RR427

Library of Congress Control Number: 2015945045

ISBN: 978-0-8330-8512-2

Published by the RAND Corporation, Santa Monica, Calif.

© Copyright 2015 RAND Corporation

RAND® is a registered trademark.

Support RAND

Make a tax-deductible charitable contribution at
www.rand.org/giving/contribute

www.rand.org

Preface

For the past 20 years, the U.S. Air Force has operated in contingency operations with relative impunity from enemy attacks. However, as U.S. national security policy focuses on ensuring U.S. presence in the Pacific region, the Air Force must be prepared to deploy and operate its forces in environments where air bases could be subject to attack. A near-peer power in this region has significant ballistic and cruise missile capabilities that can damage U.S. and allied air bases. Ensuring resilient operational capabilities in denied environments will likely require a mix of strategies, including active defense, dispersal of operating locations and forces, some hardening of base facilities to protect aircraft and other high-value assets, and some combat support recovery functions (such as airfield damage repair capability).

The combat support requirements and resource investments needed to support operations in denied environments, to include dispersed operations and other mitigation strategies, have not been comprehensively examined across a range of scenarios. In 2011, the Air Force asked RAND Project AIR FORCE (PAF) to conduct such an analysis. The work was sponsored by AF/A4/7 and AF/A3/5 in fiscal year (FY) 2012, by Air Force Materiel Command and the Policy branch of the Office of the Secretary of Defense in FY 2013, and by the Pacific Air Forces in FY 2013–2014, and was conducted within the Resource Management Program of RAND Project AIR FORCE. The research focused on four main questions:

1. How does dispersed basing affect combat support resource requirements, and how many operating locations can be supported?
2. Where should combat support resources be stored and maintained to enable rapid deployment and employment of forces in the Pacific theater?
3. How vulnerable are U.S. air bases in denied environments, given current and planned capabilities?
4. What is the right mix and level of investment in active and passive defense materiel solutions to ensure resilience against a range of adversary kinetic attack strategies?

This report documents the modeling framework that PAF developed to analyze support for combat operations in denied environments. The framework consists of two models developed as part of prior PAF research and two new models developed specifically for this analysis. Together these models help illuminate combat support requirements, vulnerabilities, resiliency, and capability trade-offs and enable decisions concerning force posture, current and future investments, and theater-shaping strategies.

While this report illustrates model outputs using a notional scenario, the focus is on modeling capabilities rather than on results. A separate report that details PAF's findings for a specific scenario and operational plan is forthcoming.

This report should be of interest to analysts and policymakers who are examining the vulnerabilities, damage expectancies, and cost-effectiveness of mitigation strategies associated with attacks on air bases.

RAND Project AIR FORCE

RAND Project AIR FORCE (PAF), a division of the RAND Corporation, is the U.S. Air Force's federally funded research and development center for studies and analyses. PAF provides the Air Force with independent analyses of policy alternatives affecting the development, employment, combat readiness, and support of current and future air, space, and cyber forces. Research is conducted in four programs: Force Modernization and Employment; Manpower, Personnel, and Training; Resource Management; and Strategy and Doctrine. The research reported here was prepared under contract FA7014-06-C-0001.

Additional information about PAF is available on our website: http://www.rand.org/paf/

Contents

Figures

Tables

Summary

For the past two decades, the U.S. Air Force has operated with impunity from air bases that have been relatively safe from attack. This may become more challenging in the future, as U.S. security policy places greater emphasis on ensuring U.S. presence in the Pacific. The "rebalance to the Pacific" could expose U.S. and allied air bases to significant ballistic and cruise missiles from a near-peer power. Ensuring resilient combat operations in denied environments (CODE) will likely require a mix of strategies, including active defense, dispersed operations, some hardening, and some combat support recovery functions (such as airfield damage repair [ADR] capabilities). These measures will have substantial impacts on combat support materiel requirements and logistics; force posturing; base infrastructure requirements, including resources needed to mitigate threats; and relationships between the United States and its allies. Yet, thus far, there has been no comprehensive analysis of these impacts and how the Air Force should manage them.

This report presents an analytic framework, developed by RAND Project AIR FORCE, to help Air Force leaders think through the challenges involved in CODE.[1] The framework was developed as part of an ongoing project commissioned by AF/A4/7 and AF/A3/5 in fiscal year (FY) 2012, by Air Force Materiel Command (AFMC) and the Policy branch of the Office of the Secretary of Defense (OSD) in FY 2013, and by the Pacific Air Forces (PACAF) in FY 2013–2014. The purpose of the research was to address four main questions:

1. How does dispersed basing affect combat support resource requirements, and how many operating locations can be supported?
2. Where should combat support resources be stored and maintained to enable rapid deployment and employment of forces in the Pacific theater?
3. How vulnerable are U.S. air bases in denied environments, given current and planned capabilities?
4. What is the right mix and level of investment in active and passive defense materiel solutions to ensure resilience against a range of adversary kinetic attack strategies?

Analytic Framework Overview

To answer these questions, PAF constructed a suite of models and tools, depicted in Figure S.1. The framework allows analysts to test a wide range of assumptions and courses of action, including defense planning guidance scenarios, adversary arsenal (e.g., quiver size and targeting apportionment), major command operational plans (e.g., base beddown and sortie generation requirements), and combat support mitigation strategies (e.g., basing dispersal options and

---

[1] A separate report that details PAF's findings for a specific scenario and operational plan is forthcoming.

repair/recovery capabilities). The major outputs include detailed lists of combat support requirements to support specific basing postures in specific scenarios; optimal locations to preposition war reserve materiel (WRM); optimal investments in active and defense resources; and, most important, the effects of all the above factors on Blue's ability to generate sorties in denied environments.

Figure S.1. Analytic Framework for Overall CODE Analyses

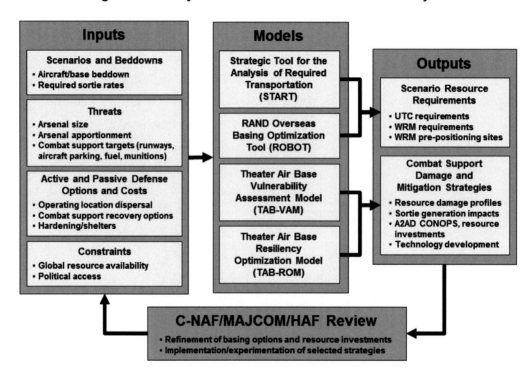

The heart of the framework is a set of four models, shown in the center of Figure S.1. START and ROBOT were developed as part of prior PAF work and adapted for use in the CODE analysis. TAB-VAM and TAB-ROM were developed specifically for the CODE analysis.

The major inputs, outputs, and applications of each are summarized in the sections below.

Strategic Tool for the Analysis of Required Transportation

As the Air Force considers distributed basing strategies to mitigate anti-access/area-denial (A2AD) threats, a primary question is what combat support resources are required to support dispersal. Historically, combat support (which includes civil engineering, communications, security forces, maintenance, services, munitions, and other functions) has dominated the footprint at operating locations.[2]

---

[2] Lynch et al., 2005.

START is an Excel-based model that estimates manpower and equipment requirements, measured in Unit Type Codes (UTCs), to support a given basing posture in a combat scenario. START was developed by PAF in 2004 as part of a larger examination of Air Force deployment requirements.[3] The model takes air order of battle–level inputs (e.g., aircraft beddown, base infrastructure, sortie requirements, and level of threat) for a single operating location, quantifies the logistics support requirements, and generates a list of UTCs to support an operation at that location. It then sums results across all bases to produce theater-wide requirements. START estimates UTCs for core capabilities in the following functional areas: aviation and maintenance, aerial port operations, civil engineering, bare-base support, munitions, fuels mobility support equipment, deployed communications, force protection, medical support, and general-purpose vehicles. These capabilities constitute the vast majority of the mass and volume of materiel that must be at a site to initiate and sustain operations.

In the CODE analysis, we use START to define UTC requirements for a range of dispersed basing options. We compare these results to existing resources to determine whether a given beddown is supportable and which UTCs are limiting factors.

RAND Overseas Basing Optimization Tool

Having identified materiel requirements to support operating locations in a given scenario, we next consider where non-unit combat support resources (i.e., those that do not deploy with the unit) should be stored and maintained to enable rapid deployment and employment of the force.

ROBOT is a mixed integer program, developed as part of prior PAF research, that identifies the least-cost allocation of war reserve materiel (i.e., non-unit) resources among existing and potential storage locations and determines a transportation network necessary to satisfy a set of time-phased operational requirements.[4] ROBOT uses START data to generate combat support requirements needed to support the contingency operation. It explicitly models transportation constraints, facility constraints, and time-phased demand for commodities at forward locations. The output is a network that connects a set of forward support locations (FSLs) with forward operating locations (FOLs). ROBOT finds the lowest-cost FSL network that satisfies operational requirements over a given time horizon. That is, the costs of conducting training and deterrence missions are minimized, while the solution set is constrained to have the storage and throughput required to meet contingency scenarios should deterrence fail.

In the CODE analysis, ROBOT can be used to evaluate the costs and risks of prepositioning postures to meet a range of scenarios on a regional or global scale. Prior PAF analyses suggest

---

[3] Snyder and Mills, 2004.

[4] Amouzegar et al., 2006.

xiii

that a globally managed posture can support a robust set of contingencies at lower cost and with lower risk (e.g., from network disruptions) than a regionally managed posture.[5]

Theater Air Base Vulnerability Assessment Model

The next questions to consider are how operational plans perform under attack and how threat mitigation options affect operational performance and cost. PAF developed TAB-VAM, a Monte Carlo simulation model, to analyze the complex trade-offs among basing strategies and threat mitigation options. The model allows the user to assess and compare a wide range of scenarios, aircraft beddowns, base recovery capabilities, infrastructure investments, passive and active missile defense options, and concepts of operations (CONOPS).

The major inputs and outputs of TAB-VAM are shown in Figure S.2. For a given model run, the user specifies the air bases to be considered within the theater of operations; the resources at each base (including runways, aircraft parking, fuel storage, active and passive missile defense, and damage repair capabilities); a time-phased aircraft beddown that includes locations and target sortie rates for all Blue aircraft in the scenario; and an enemy attack strategy that includes the arsenal of ballistic and cruise missiles and the relative allocation of each at different bases and airbase targets.

Figure S.2. Major TAB-VAM Inputs and Outputs

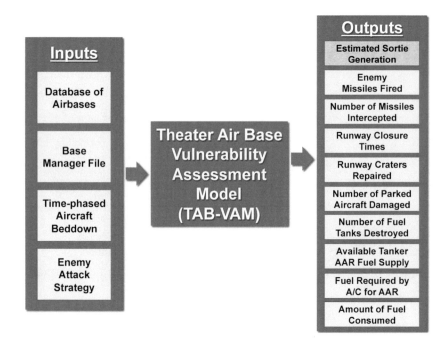

---

[5] McGarvey et al., 2010.

Based on these inputs, TAB-VAM simulates Red missile attacks; Blue active and passive missile defense; the effects of attacks on runways, parked aircraft, and fuel storage; and the effectiveness of Blue repair capabilities and threat mitigation options (e.g., fuel bladders, hardened aircraft shelters, or ADR teams). The model can be expanded to examine a wider range of resources and investment options. The principal output and measure of performance for a given model run is the percentage of planned sorties generated at each base on each day of the conflict.[6] We use sortie generation as the primary metric because it focuses attention on the resource investments and force planning decisions that *enable* combat operations in denied environments. We do not model the operational effectiveness of sorties once they are launched.

TAB-VAM is a powerful tool for understanding the challenges and complexities of air operations in denied environments. The model provides a robust, detailed, multilayered representation of the factors that can affect Blue performance. By allowing the user to vary many dimensions of Blue and Red operations and resources, TAB-VAM provides an assessment of uncertainties and risks. It also allows for a robust exploration of trade-offs among different basing strategies and resource investments. In the CODE analysis, we use TAB-VAM to examine the vulnerability of specific basing postures in a given scenario. We also use TAB-VAM to understand the trade-offs among investments—or combinations of investments—in missile defense, ADR, fuel infrastructure, and hardened aircraft shelters. When combined with TAB-ROM, its companion model, TAB-VAM can analyze various mixes of investments at different budget levels and where resources should be utilized. Thus, it can enable robust force planning against a variety of near-term and far-term threat scenarios.

Theater Air Base Resiliency Optimization Model

PAF developed TAB-ROM as a companion to TAB-VAM. TAB-ROM searches the entire space of user-defined enemy attack strategies for a given scenario and finds the most cost-effective way to improve Blue sortie generation through investments in active missile defense, hardened aircraft shelters, fuel storage, and/or ADR. As with TAB-VAM, TAB-ROM can be expanded to examine a wider range of resources and investment options. TAB-ROM interacts with TAB-VAM to compute the sortie generation impacts of making various investments. In the CODE analysis, we compare TAB-ROM runs at various budget levels to determine how much should be spent (and on what) to achieve a given sortie generation goal and to identify the point at which further spending yields diminishing returns.

Conclusions

This report describes the suite of models that PAF is using to analyze support for combat operations in denied environments. We use these models to examine a variety of questions of

---

[6] The user may choose to examine other metrics, such as the number of damaged aircraft.

interest to force planners. For example, the CODE analysis shows the implications of dispersal on combat support resource requirements (START). The CODE analysis also assesses air base vulnerability in a given scenario and shows the advantages of specific investments—and combinations of investments—in base infrastructure and damage repair capabilities (TAB-VAM and TAB-ROM).

By providing insights into combat support requirements, vulnerabilities, resiliency, and capability trade-offs, the modeling framework can help inform the development of operational and support CONOPS in denied environments, current and future investment decisions, area of responsibility basing strategies, discussion within the Secretary of Defense's Management Action Group, and Air Force advocacy for the Quadrennial Defense Review.

Further modeling development is in progress to extend the breadth and fidelity of PAF's analytic framework. These areas include

- improving the modeling fidelity of active missile defense assets, including expanded interceptor inventory and the vulnerability of launch platforms to adversary attack
- expanding the models' visibility into Blue munitions, including factors such as theater inventory, alternative munitions storage CONOPS, and assessing the vulnerability of munitions storage areas
- broadening TAB-VAM's assessments of fuel infrastructure to include attacks on the broader supply chain of intra-theater fuel transportation, as well as receipt and distribution at individual operating locations
- extending the assessment of attacks to include the broader fuel supply chain, to include fuel receipt and distribution
- assessing the impacts of adversary attacks on maintenance personnel and repair facilities on sortie generation capability
- including the role of cyber attacks on the disruption of combat support.

Many of these additional modeling features will draw on expertise from the Joint community, given that factors such as fuel, electricity, materiel delivery, and munitions influence more than just the Air Force.

Acknowledgments

This work would not have been possible without the support from many people from a wide variety of communities within the Air Force and OSD.[7] First, we thank Generals Herbert Carlisle, PACAF/CC, and Janet Wolfenbarger, AFMC/CC, for sponsoring this work. Gen Carlisle was briefed several times on this work, and he selected specific results to brief senior officials within OSD on the Air Force strategy that was developed to enable the "rebalance [in national attention] to the Pacific."

Many at PACAF informed the development of the analytic framework that was used in developing and evaluating CONOPS, investments, and basing strategies. Lt Gen Ted Kresge, PACAF/CV, provided specific guidance on the required content of the CODE framework. Other senior members of the PACAF staff directly contributed to this analysis, including: Brig Gen Steven Basham, PACAF/A5/8, Brig Gen Scott West, PACAF/A3/5/8, Col Carl Bosworth, PACAF/A7, Brig Gen Carl Buhler, PACAF/A4, Brig Gen Pat Malackowski, PACAF/A8/9, Ron Kennedy, PACAF/A8/9, Jim Silva, PACAF/A4, Col Danny Wolf, PACAF/A3X, Kenneth Dorner, PACAF/A3XI, Col Kevin Sampels, PACAF/A4P, Russell Grunch, PACAF/A4PP, Maj Chad Sitzmann, PACAF/A4PX, Elaine Ayers, PACAF/A4PX, MSgt David Martin, PACAF/A4PX, Maj Douglas Hellinger, PACAF/A5XS, John Ahern, PACAF/A7XX, Col Mark Bednar, PACAF/A7X, Mark Leonard, PACAF/A7O, Maj Matthew Anderson, PACAF/A7XZ, John Trifonovitch, PACAF/A9A, Lt Col Michael Artelli, PACAF/A9A, Christopher Pitcher, MITRE, and Gilbert Loomis, PACAF/A2X.

Col Mark Harysch, PACAF 613 AOC/JTF 519 JFACC LNO, paid particular attention to the CODE analytic framework and was instrumental in having the CODE team participate in the PACOM IAMD Wargame IV, supported by the Naval War College, and held at PACAF in January 2013.

Col Hugh Hanlon, PACAF/A8X, chaired three worldwide colonel-level reviews of CODE and facilitated Air Force–wide understanding and support for the development and use of the CODE analytic framework. Lt Col Ray Alves, PACAF/A8XP, became the CODE focal point in PACAF on Colonel Hanlon's departure from PACAF and has carried on the leadership role for shaping CODE analyses to meet the needs of PACAF senior leaders. Many on the PACAF/A8X staff worked with us in shaping the CODE framework and in developing inputs for Gen Carlisle's briefing to the DEPSECDEF. Maj Matt Forner, Robert Craven, Maj Todd Larson, and Jeff LeVault, Booz-Allen, deserve special recognition.

We are grateful to David Ochmanek and Col Chris Niemi from OSD-Policy for their support of CODE sensitivity analyses. Col Niemi hosted weekly meetings to ensure that CODE

---

[7] All ranks and offices are current at the time of the research.

assumptions and analyses were sufficiently robust to provide inputs to the Program Budget Review and Quadrennial Defense Review.

From AFMC, we would like to thank Brig Gen Stephen Denker, AFMC/A8/9, Col John Long, AFMC A8/9, Col Chris Froehlich, AFMC/A8X, Bill Santiago, AFMC/A8X, and Col Randall Gilhart, AFMC/A4A.

From AMC, we thank Col Michael Peet, AMC/A5X.

At the Air Force Civil Engineering Center and Air Force Research Labs we thank: Col Michael Mendoza, AFCEC/CX, Craig Rutland, AFCEC/COSC, Lance Filler, AFCEC/CXXE, Gregory Cummings, AFCEC/CXX, Eugene Kensky, AFMC AFRL/RXQEM, and Dr. Mike Hammons, AFRL.

At the Air Staff, we thank, Lt Gen Judith Fedder, AF/A4/7, Maj Gen Theresa Carter, AF/A7C, Maj Gen John Cooper, AF/A4L, Mark Correll, AF/A7, John Weida, AF/A5X, Brig Gen (s) John Cherrey, AF/A5XS, Col Jordan Thomas, AFA5XS, Lt Col Nathan Mead, AF/A5XS, Col Rich Gannon, AF/A9F, Ray Miller, AF/A9FC, Larry Parthum, AF/A9FC, Laine Krat, AF/A4LX, Col Michael Kozak, AF/A7CX, Col Ed Oshiba, AF/A7CI, and Lt Col Matt Brennan, AF/A7CIP. A special thanks goes to Dr. Carl Rehberg, AF/A8XC-APC (Asia Pacific Cell), for moving this work forward.

At RAND, we thank Ted Harshberger, Laura Baldwin, Marygail Brauner, Jim Chow, Paul Davis, John Drew, Sarah Evans, Michael Kennedy, Sherrill Lingel, Ron McGarvey, Pat Mills, and Chris Mouton for their support and guidance of this effort.

Abbreviations

A2AD	anti-access/area denial
AAR	air-to-air refueling
A/C	aircraft
ACC	Air Combat Command
ADR	airfield damage repair
AEF	Air Expeditionary Force
AFCEC	Air Force Civil Engineering Center
AFMC	Air Force Materiel Command
AFPA	Air Force Petroleum Agency
AFRL	Air Force Research Laboratory
ALCM	air-launched cruise missile
AMC	Air Mobility Command
APF	afloat preposition fleet
ART	AEF reporting tool
BDA	battle damage assessment
BM	ballistic missile
C-NAF	Component Numbered Air Force
C2	command and control
C2ISR	command, control, intelligence, surveillance, and reconnaissance
CE	civil engineering
CEP	circular error probable
CM	cruise missile
CMD	cruise missile defense
CODE	combat operations in denied environments
CONOP	concept of operations
CS	combat support

CSAR	combat search and rescue
CSV	comma-separated value
DMAG	Deputy Secretary of Defense's Management Action Group
DoD	Department of Defense
EOD	explosive ordnance disposal
EW	electronic warfare
FABO	forward air base operations
FOC	full operating capability
FOL	forward operating location
FORCE	fuels operational readiness capability equipment
FSL	forward support location
FY	fiscal year
GA	genetic algorithm
GAMS	General Algebraic Modeling System
GLCM	ground-launched cruise missile
GUI	graphical user interface
HAF	Headquarters Air Force
HIPPO	hardened installation protection for persistent operations
HSS	high-speed sealift
IAMD	Integrated Air and Missile Defense
IDA	Institute for Defense Analyses
IOC	initial operating capability
IRBM	intermediate-range ballistic missile
JTF NA	Joint Task Force Noble Anvil
LP	linear programming
MAJCOM	major command
MINLP	mixed integer nonlinear programming
MIP	mixed integer programming

MOB	main operating base
MOG	maximum on ground
MOS	minimum operating surface
MRBM	medium-range ballistic missile
MX	maintenance
NBC	nuclear, biological, and chemical
OEF	Operation Enduring Freedom
OEPP	Operational Energy Plans and Programs
OIF	Operation Iraqi Freedom
OSD	Office of the Secretary of Defense
PACAF	Pacific Air Forces
PAF	Project AIR FORCE
QDR	Quadrennial Defense Review
RED HORSE	Rapid Engineer Deployable Heavy Operations Repair Squadron
ROBOT	RAND Overseas Basing Optimization Tool
SOF	Special Operations Forces
SRBM	short-range ballistic missile
START	Strategic Tool for the Analysis of Required Transportation
TAB-ROM	Theater Air Base Resiliency Optimization Model
TAB-VAM	Theater Air Base Vulnerability Assessment Model
THAAD	Terminal High Altitude Air Defense
TPFDD	time phased force deployment data
USTRANSCOM	United States Transportation Command
UTC	unit type code
WRM	war reserve materiel
XML	Extensible Markup Language

1. Introduction

For many years, the U.S. Air Force has operated out of air bases that have largely been safe havens and conducted operations with relative impunity from attack.[8] In the Pacific theater, a near-peer power has significant ballistic and cruise missile capabilities that can damage air bases where U.S. and allied air forces might operate. To support the national objective of ensuring U.S. presence in the Pacific region (sometimes referred to as the "rebalance to the Pacific"), the Air Force must examine how to deploy and operate its forces in environments where air bases could be subject to attack. Doing so requires the development of concepts of operations (CONOPS), investments, and basing strategies capable of absorbing such attacks and continuing operations.

One way to complicate potential adversary targeting is to disperse U.S. forces across a larger number of operating locations. Another way to protect forces under attack is to harden facilities to shelter aircraft and other high-value assets. Ensuring resilient operating capabilities in denied environments will likely require a mix of strategies, including active defense, dispersed operations, some hardening, and some combat support recovery functions (such as airfield damage repair [ADR] capabilities).

These measures will have substantial impacts on combat support materiel requirements and logistics; force posturing; base infrastructure requirements, including resources needed to mitigate threats; and relationships between the United States and its allies. Yet, the combat support resource requirements and mitigation options needed to support operations in denied environments have not been comprehensively examined across a range of scenarios. As policymakers formulate CONOPS, investment priorities, and theater-shaping strategies to counter anti-access/area-denial (A2AD)[9] challenges, several important questions must be addressed:

1. How does dispersed basing affect combat support resource requirements, and how many operating locations can be supported?
2. Where should combat support resources be stored and maintained to enable rapid deployment and employment of forces in the Pacific theater?
3. How vulnerable are U.S. air bases in denied environments, given current and planned capabilities?
4. What is the right mix and level of investment in active and passive defense materiel solutions to ensure resilience against a range of adversary kinetic attack strategies?

---

[8] Even in recent operations, however, air bases have not been entirely immune from attack. The Taliban attack on Camp Bastion in Afghanistan in September 2012 successfully targeted parked aircraft, refueling stations, aircraft hangars, and personnel.

[9] *Anti-access* refers to the ability to gain access to the area or theater of engagement. *Area denial* refers to the ability to employ within the areas or theater of engagement.

The purpose of this report is to present an analytic framework for evaluating combat support requirements, vulnerabilities, resiliency, and potential mitigation strategies to support combat operations in denied environments (CODE). The framework was developed as part of a RAND Project AIR FORCE (PAF) project sponsored by AF/A4/7 and AF/A3/5 in fiscal year (FY) 2012, by Air Force Materiel Command (AFMC) and the Policy branch of the Office of the Secretary of Defense (OSD) in FY 2013, and by the Pacific Air Forces (PACAF) in FY 2013–2014. Using this framework, analysts can assess combat support network design vulnerabilities against kinetic attacks by near-peer or regional powers and the costs, effectiveness, and risks of alternative strategies to mitigate potential adversary A2AD capabilities. By providing insights into combat support requirements, vulnerabilities, resiliency, and capability trade-offs, the modeling framework can help inform the development of operational and support CONOPS in denied environments, current and future investment decisions, area of responsibility basing strategies, discussion within the Deputy Secretary of Defense's Management Action Group (DMAG), and Air Force advocacy for the Quadrennial Defense Review (QDR).

RAND has a long history of investigating issues relating to A2AD and base resilience. During the Cold War, RAND investigated the resilience of bases under attack from the Soviet threat. Tools supporting these efforts tended to be highly detailed, such as the Theater Simulation of Airfield Resources model, which used factors such as manning levels in maintenance squadrons and spares availability to sustain sortie generation capability.[10] More recently, RAND has shifted focus to examine broader U.S. defense postures, focusing on strategic factors such as political access, deterrence, and contingency responsiveness.[11] The analytic framework presented in this document strives to bridge the gap between specific investment options affecting base resilience and their role in supporting theater-wide basing postures.

Analytic Framework Overview

Figure 1.1 provides an overview of the analytic framework and its major inputs and outputs. The framework allows users the flexibility to test various assumptions for a range of chosen scenarios. As shown on the left side, the user inputs information about the operational plan (e.g., base beddown and sortie generation requirements), adversary arsenal (e.g., quiver size and targeting apportionment), combat support mitigation strategies (e.g., basing dispersal options and repair/recovery capabilities), and any combat support constraints (e.g., worldwide inventory and host nation support). These inputs are utilized by a suite of four models, developed by PAF, which roughly correspond to the four questions listed above:

- The *Strategic Tool for the Analysis of Required Transportation* (START) is a rule-based spreadsheet model that translates specified operational capability at multiple locations

---

[10] Emerson, 1992.

[11] Morgan, 2012; Lostumbo et al., 2013.

into a list of Unit Type Codes (UTCs) needed to generate theater-wide capability. While developed as part of prior PAF research, in the CODE analysis it is used to determine how combat support resource limitations may restrict operating location strategies.

- The *RAND Overseas Basing Optimization Tool* (ROBOT) is a mixed integer program that identifies the least-cost allocation of war reserve materiel (WRM) resources (i.e., non-unit combat support) among existing and potential storage locations and determines a transportation network necessary to satisfy a set of time-phased operational requirements. Also developed as part of prior PAF research, in the CODE analysis it is used to determine where WRM should be stored and maintained to support rapid deployment and employment of the force.

- The *Theater Air Base Vulnerability Assessment Model* (TAB-VAM) is a Monte Carlo simulation developed in FYs 2012–2014 that simulates an A2AD attack on air bases and then estimates the damage to each base and the effects on theater-wide sortie generation capability. It allows the user to test the effects of various aircraft beddown postures, investments in air base resources (e.g., ADR, aircraft parking shelters, fuel storage facilities, and active missile defense), and assumptions about enemy missile threats.

- The *Theater Air Base Resiliency Optimization Model* (TAB-ROM) is a genetic algorithm developed in FYs 2012–2014 that works with TAB-VAM to identify optimal U.S. and allied investment strategies to mitigate A2AD threats and to maximize sortie generation within a given budget.

Figure 1.1. Analytic Framework for Overall CODE Analyses

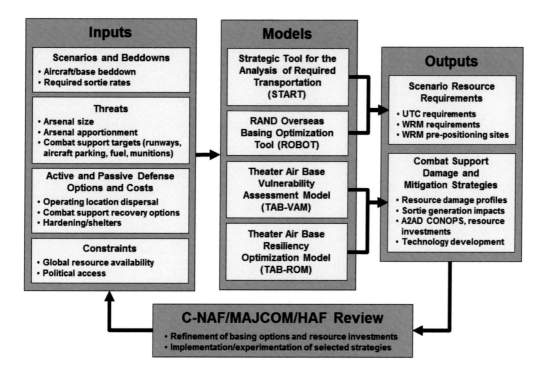

Organization of This Report

This report describes the four principal models that constitute the analytic framework that PAF is using to analyze combat support operations in denied environments. Chapters Two through Five describe START, ROBOT, TAB-VAM, and TAB-ROM, respectively. In each case, we use a notional scenario (described in the appendix) to illustrate the inputs, modeling, and outputs. Chapter Six discusses how the framework as a whole can be used to help develop operational and support CONOPS and investment strategies. While this document provides notional "results" to illustrate model outputs, the focus is *modeling capability* rather than findings.[12]

---

[12] A separate report that details PAF's findings for a specific scenario and operational plan is forthcoming.

2. Strategic Tool for the Analysis of Required Transportation

As the Air Force considers distributed basing strategies to mitigate A2AD threats, a primary question is what combat support resources are required to support dispersed operations. Historically, combat support (which includes civil engineering, communications, security forces, maintenance, services, munitions, and other functions) has dominated the footprint at operating locations. As shown in Figure 2.1, aviation units and their associated maintenance functions accounted for between 9 and 20 percent of tonnage moved to operating locations in Joint Task Force Noble Anvil (JTF NA), Operating Enduring Freedom (OEF), and Operation Iraqi Freedom (OIF). Aerial port equipment (which includes Tactical Airlift Control Element and associated items) accounted for between 6 and 8 percent. Combat support accounted for between 74 and 84 percent, making it a potential limiting factor in the United States' ability to operate overseas.[13]

Figure 2.1. Tonnage Moved to Operating Locations in Recent Operations

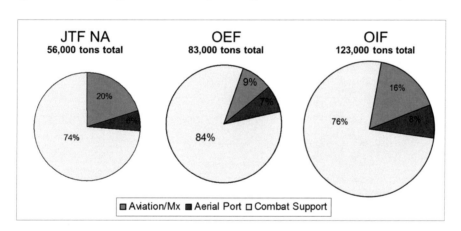

Planners must ensure enough resources are available to support dispersal to additional operating locations. This chapter describes START, an Excel-based model that estimates manpower and equipment requirements, measured in UTCs, to support a given basing posture in a combat scenario. START was developed by PAF in 2004 as part of a larger analysis of Air Force deployment requirements.[14]

---

[13] Lynch et al., 2005.

[14] START has multiple applications beyond those utilized in the CODE analysis. In addition to estimating UTC requirements, it can estimate the movements required to achieve operating capability at each location. Thus, it can generate a first approximation of Time-Phased Force Deployment Data (TPFDD) within minutes without logistics planners and without the time and security concerns associated with the current TPFDD-building process. Moreover, START can rapidly evaluate requirements for dozens of scenarios, thereby assisting with capability-based planning. For a comprehensive description of START, see Snyder and Mills, 2004.

The model takes air order of battle–level inputs (e.g., aircraft beddown, base infrastructure, sortie requirements, and level of threat) for a single operating location, quantifies the logistics support requirements, and generates a list of UTCs to support an operation at that location. It then sums results across all bases to produce theater-wide requirements. START estimates UTCs for core capabilities in the following functional areas: aviation and maintenance, aerial port operations, civil engineering, bare-base support, munitions, fuels mobility support equipment, deployed communications, force protection, medical support, and general-purpose vehicles. These capabilities constitute the vast majority of the mass and volume of materiel that must be at a site to initiate and sustain operations.[15]

The user can determine the aggregate requirement to support an overall contingency by adding the requirements at each location and subtracting theater-level efficiencies and resources already in theater. By comparing the START output to existing resources, the user can determine whether a given beddown is supportable and which UTCs are limiting factors. START can also be used to identify optimal prepositioning locations as part of a ROBOT analysis (as discussed in the next chapter).

This chapter describes the major inputs and outputs of START as it is employed in PAF's CODE analysis and illustrates a model run using the notional scenario described in the appendix.

Inputs and Outputs

The factors that principally drive materiel and manpower needs for an operational capability are base infrastructure, the level of threat to which the location is exposed, aircraft beddown, and sortie rate. We describe these inputs and outputs below.

Base Infrastructure

Materiel needs are largely driven by the amount of existing infrastructure at a base and the number of aircraft that can be simultaneously serviced on the ramp, whether refueling or loading/unloading cargo (expressed as Maximum on Ground [MOG] capability).

For each location, the user specifies either a *bare* or *established* base and whether the base has any Air Force presence. A bare base has a usable runway, taxiway, parking areas, and a source of water that can be made potable. The baseline assumption is that anything needed for operations must be supplied. In the CODE analysis, as a default, we assume that heavy construction (e.g., building or runway construction) is not required. If needed, the user can specify this requirement, and the appropriate Rapid Engineer Deployable Heavy Operations Repair Squadron (RED HORSE) teams are added to the movement requirements.

---

[15] UTCs not included in the model (e.g., consumables such as food and fuel) are generally deployed only under special circumstances or are comparatively light.

Alternatively, an established base refers to a main operating base (MOB), international airport, allied-country military base, and so forth. The user determines the additional infrastructure needed to achieve the desired capability. Examples of infrastructure considered by START include whether a new airframe will be introduced to the site and whether additional billeting, communications, fuels equipment, medical facilities, and force protection are required. If heavy construction is needed, the user can select whether it is horizontal (e.g., ramps, runways) or vertical (e.g., buildings). This range of options allows the user flexibility to tailor the characteristics of a deployed location.

The base layout and topography can also substantially affect requirements for functional areas. START estimates the requirements for a given operational capability using a typical deployed base layout and topography. As the model is designed for strategic, not tactical, use, it keeps these inputs as general as possible.

START allows the user to specify whether the calculation is for initial operational capability (IOC) or full operating capability (FOC). FOC means IOC plus maintenance equipment for operations beyond 30 days and munitions for operations up to seven days.[16]

Threat Level

The threat level at a given base determines the requirements for force protection, explosive ordnance disposal (EOD), and medical support. The user specifies whether each base is subject to *conventional* and/or *nuclear, biological, and chemical (NBC)* threats and enters a rating from "Non" to "High." The conventional threat level measures the vulnerability of the base to ground attack and is used to determine the amount of force protection needed. It does not include capabilities that are not organic to the Air Force, such as Patriot missile batteries or heavy ground troops. The NBC threat level measures the likelihood of attack by nonconventional weapons. It determines needs in the areas of medical support and engineering readiness.

Aircraft Beddown and Sortie Rate

The user specifies the type, number, and sortie rate of aircraft bedded down at each operating location in the scenario. Most aircraft are listed and grouped as fighters and attack aircraft; Special Operations Forces (SOF) aircraft; bombers; mobility aircraft; and command, control, intelligence, surveillance, and reconnaissance (C2ISR) assets. These inputs drive materiel needs in areas such as aviation, maintenance, aerial port operations, munitions, and munitions handling.

---

[16] Although not utilized in the CODE analysis, the user can also specify that a base is used for theater operations, such as a regional hospital or Air Operations Center. START modifies UTC requirements accordingly. Other START capabilities not currently utilized in the CODE analysis include estimates of munitions storage and basic expeditionary airfield resource (BEAR) storage requirements.

Total Base Population

START uses the total base population to determine the demand for many support UTCs. The base population is initially estimated from the number of aircraft bedded down, using bare-base planning factors. These planning factors give a range of anticipated base population as a function of the number and size of the aircraft bedded down at the site, taking a conservative, upper estimate. This total base population principally drives materiel and personnel needs in civil engineering, bare-base support, medical services, and communications.

Adding UTCs to support the base population further increases that population. Therefore, several model iterations are needed to reach a final population. The total base population typically converges after two iterations, and so START executes this process twice.

Outputs

Each START run generates a "Base List" worksheet, which shows the UTC requirements for a given base. The table includes the UTC, UTC title, quantity required, function abbreviation, function name, weight per UTC in tons, personnel per UTC, total weight in tons, and total personnel. In the CODE analysis, the user manually adds the UTC requirements for multiple bases to reach the total requirements for a given force posture.

Illustrative Model Run

To illustrate how START is utilized in the CODE analysis, we generated UTC requirements for the notional Consolidated and Dispersed Beddowns described in the appendix. The notional scenario describes two U.S. and allied beddowns (Consolidated and Dispersed) at multiple air bases on a series of island chains at various distances from an adversary mainland. See the appendix for details.

Inputs

We entered information about each base using a series of tabs, illustrated in Figures 2.2 and 2.3. In these examples, we show inputs for Base 6, a bare base on Island C. This base is not used in the Consolidated Beddown. In the Dispersed Beddown, eight tankers are moved to Base 6, which is outside the medium-range ballistic missile (MRBM) threat ring.

First, as shown in Figure 2.2, we entered information about the type of base, the resources required, the threat level, and the MOG. For the purposes of this example, we assume that Base 6 is a bare base with no Air Force presence. It has a usable runway, taxiway, parking area, and source of water that can be made potable. Anything further needed for operations (e.g., billeting, medical, vehicles) must be supplied, as indicated by the checkmarks. The amount of resources in these areas is based on the number of personnel and aircraft that will be located at this base.[17]

---

[17] Although not utilized in this scenario, START can also model munitions inventory at each base.

The conventional and NBC threat levels are rated "high" because the base is still not far from the adversary and within the intermediate-range ballistic missile (IRBM) threat ring. These selections will drive equipment and manpower needs in the areas of force protection, EOD, and medical support. We assume there is no billeting available for deploying personnel and that there are no additional personnel beyond those estimated from UTC requirements. In the MOG field, we specify that two Air Force mobility aircraft can be serviced on the ramp, whether refueling or loading or unloading cargo simultaneously, and that the base does not service commercial aircraft.[18]

Figure 2.2. START Base Input Screen

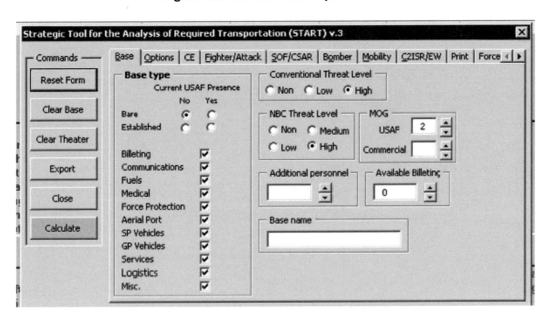

Next, the user specifies the type, number, and sortie rate of aircraft at the base. These inputs drive UTCs in deployed communications, fuels mobility support equipment, aerial port requirements, and several bare base planning factors (e.g., personnel estimates, base support, civil engineering, medical). Figure 2.3 shows the inputs for Base 6 in the Dispersed Beddown. There are eight KC-135s each with a sortie rate of one per day. In a more extensive scenario, the user would be able to specify special operations; combat search and rescue; bombers; mobility; command and control; intelligence, surveillance, and reconnaissance; or electronic warfare aircraft, as the case may require. These options are represented by the tabs on the top of Figure 2.3.

---

[18] Some basic equipment is required to support aerial port operations (APO) at a bare base, regardless of cargo flow (e.g., lights, special-purpose trucks). However, material handling equipment and manpower requirements are driven by MOG. When transport aircraft are included in a base beddown, we set the MOG to 4 to reflect the greater demand on APO. Other APO resources are already captured in the general-purpose vehicles section of the model and therefore not included here.

Figure 2.3. START Aircraft Input Screen

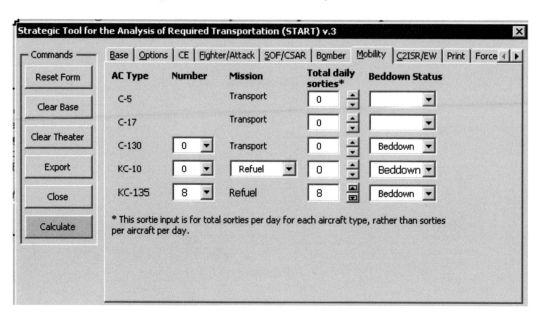

Outputs

The primary output of START is a list of UTCs for each base, a portion of which is shown in Table 2.1. In this example, START indicates that Base 6 (in the Dispersed Beddown) requires 430 UTCs which includes 1,872 tons of equipment and 870 people.

Table 2.1. UTC Requirements for Base 6 (Dispersed Beddown) (Detail)

UTC	TITLE		QUANTITY	FUNCTION	NAME	WT	PERSONNEL	TOTAL WT	TOTAL PERS
3YC4R	ARS 04 KC135R	FW1	1	ARS	Air Refueling Avia	0	18	0.0	18
3YCLR	ARS 04 KC135R	LD	1	ARS	Air Refueling Avia	6.3	28	6.3	28
4F9AL	EN FUEL SYS MX EQUIP SUP SET		1	EN	Engineering	0.1	0	0.1	0
4F9ED	EN BD/SUST FOLLOW-ON COMM EQ		1	EN	Engineering	0.8	0	0.8	0
4F9EE	EN BDOWN/SUST PEST MGT SUP EQ		1	EN	Engineering	0.6	0	0.6	0
4F9EF	EN BD/SUST FOLLOW ON EQUIP SET		2	EN	Engineering	0.9	0	1.8	0
4F9EH	EN BD/SUST SURVEY SUP EQ SET		1	EN	Engineering	1	0	1.0	0
4F9ER	EN BD/SUST COMM LEAD EQUIP SET		1	EN	Engineering	0.8	0	0.8	0
4F9ET	EN BASIC ENG BD/SUST EQUIP SET		1	EN	Engineering	4	0	4.0	0

In this illustrative model run, we made analogous inputs for each base in the two beddowns. Thus, we can use START to calculate and compare the aggregate resource requirements (by UTC) for each of the two postures. Figure 2.4 shows the total weight required for each beddown and the percentage required for aerial port equipment, aviation units and their associated maintenance functions, and combat support resources. Consistent with the historical data shown in Figure 2.1, we found that combat support resources dominate the footprint for both beddowns. Of the tonnage that needs to be moved to support the Consolidated Beddown, 78 percent is combat support. Of the tonnage for the Dispersed Beddown, 81 percent is combat support. The

Dispersed Beddown requires 17 percent more total weight to be moved than the Consolidated Beddown (thus the circle is larger).

Figure 2.4. UTC Requirements

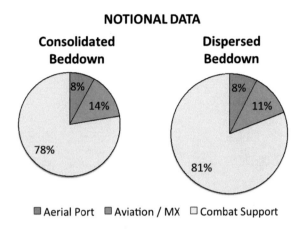

Figure 2.5 provides a more detailed comparison of UTCs for the heaviest functional areas.[19] As expected, we find that while the Dispersed Beddown increases UTC requirements overall, some functions are more sensitive to dispersal than others. Maintenance, aviation, and munitions requirements are driven more by the number of flying units and sortie generation requirements than by the number of operating locations. However, requirements for engineering, bare base support, and others increase as the number of operating locations increases. Therefore, any consideration of the operational benefits of dispersal (e.g., increased sortie generation, as discussed in Chapters Four and Five) must also consider the greater demand for combat support resources and the ability to support those demands with the expected supply. Should the analysis show that combat support resources limit dispersal, Blue may need to consider engaging Joint forces and/or allies as part of its operational plan (we discuss allied contributions to theater-wide sortie generation in Chapter Five).

---

[19] Throughout this report, we present notional results without units on the vertical axis.

Figure 2.5. UTC Requirements, by Type

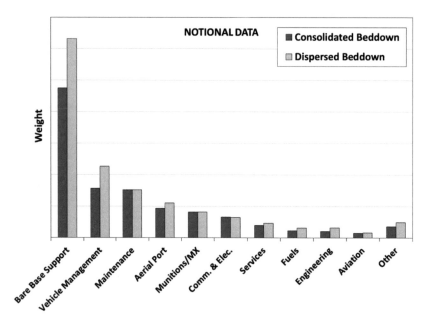

Conclusion

START was developed in 2004 as part of a broader analysis of Air Force deployment requirements. While START has features and applications beyond those described in this report, it is a principal contributor to two aspects of the CODE analysis. First, in actual model runs, we compare START results to current resources available (such as those found in the Air Expeditionary Force Reporting Tool [ART]) to help answer the question of how much dispersal can be supported. Second, we use START results as an input to ROBOT, which identifies a cost-effective strategy for prepositioning combat support materiel. We describe ROBOT in the next chapter.

12

3. RAND Overseas Basing Optimization Tool

Chapter Two discusses how the START model is used to generate materiel requirements for operating locations within a given scenario. An additional model, ROBOT, identifies the least-cost allocation of war reserve materiel (i.e., non-unit) resources among existing and potential storage locations and determines a transportation network necessary to satisfy a set of time-phased operational requirements.[20] ROBOT uses START data to generate combat support requirements needed to support the contingency operation. It explicitly models transportation constraints, facility constraints, and time-phased demand for commodities at forward locations. The output from this optimization is a network that connects a set of forward support locations (FSLs) with forward operating locations (FOLs). It allocates resources to a particular FSL and dictates the total movement of combat support resources and munitions. The model also computes the type and number of transportation vehicles required for the movement of the materiel, as well as retrograde movements.

ROBOT takes into account the objective of keeping the overall cost to a minimum, while meeting operational requirements and maximizing the support capability (e.g., reducing the time to IOC). ROBOT examines the costs and capabilities of alternative support basing options, for a constant level of performance against a variety of deployments. Thus, it can be an important tool for developing suitable programming and budgeting plans. Moreover, as illustrated at the end of this chapter, the model allows for portfolio planning options under various denied environments.

Technical Characteristics

ROBOT is a mixed integer programming (MIP) model developed using the General Algebraic Modeling System (GAMS).[21] The optimization tool is an integral part of a larger analytic model that takes into account operational scenarios, force options, and the resulting combat support resource requirements. ROBOT selects a set of FSLs that would minimize the costs of supporting a given scenario and allows for the analysis of various "what-if" questions. It also assesses the solution set in terms of resource costs for differing levels of combat support capability. Figure 3.1 depicts the overall framework of this analytic model.

---

[20] Amouzegar et al., 2006.

[21] See Brooke et al., 2003.

Figure 3.1. Analytic Framework for Determining WRM Storage and Transportation

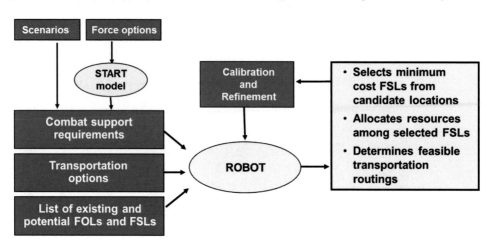

For a given deployment scenario (or a diverse set of scenarios with varying timelines) and the associated force options, the model calculates the combat support requirements, using the START tool described in Chapter Two. These requirements, along with a set of potential FSLs, FOLs, and transportation options (e.g., allowing sealift or not), serve as a starting point for ROBOT. The optimization model then selects the optimum set of FSLs that minimizes the facility operating and transportation costs associated with planned operations scheduled to take place over an extended time horizon. It also satisfies time-phased demands for combat support commodities at FOLs. The model also optimally allocates the programmed resources and commodities to those FSLs. It computes the type and the number of transportation vehicles required to move the materiel to the FOLs.

Major Model Inputs, Constraints, and Outputs

Several major constraining and contributing factors affect the capability of FSLs to support the warfighter. Our analytic framework takes each of these parameters into account in the process of selecting an optimal set of combat support locations. This section describes the major inputs, constraints, and outputs of ROBOT.[22]

Operational Scenarios

The user inputs a scenario or set of scenarios for each model run, using the combat support requirements provided by START runs. Scenarios may consist of major combat operations, continuous force presentation, small-scale humanitarian operations, and other types of military actions. The solutions returned from ROBOT are sensitive to the set of scenarios provided. This

---

[22] ROBOT adheres to typical warfighter constraints (e.g., number of available transport ships/aircraft, viable locations for WRM storage). The advanced user can examine ROBOT outputs to determine which constraints, if relaxed, would potentially reduce costs or accelerate throughput times. For a full list of parameters and constraints, see Amouzegar et al., 2006.

is why it is important to consider a broad range of potential future engagements in order to identify a robust set of facility locations.

Modes of Transportation

ROBOT allows the user to explore alternative modes of transportation, including airlift, ground transportation, and/or sealift. For example, there may be advantages to using sealift or ground transportation in place of, or in addition to, airlift. Ships have a higher hauling capacity than aircraft and can easily carry outsized or super-heavy equipment. In addition, ships do not require overflight rights from any foreign government. Aircraft are faster, but are more expensive and can carry less cargo. There are advantages to both modes, depending on whether the objective is to minimize time or cost. Allowing for alternative modes of transportation might bring some FSLs into the solution set that otherwise may have been deemed infeasible or too costly.

The user can set constraints that limit the total number of available vehicles system-wide, utilization rates and throughput, and the total vehicles available for loading at each FSL. The transportation parameters include

- transportation modes available at each FSL at the beginning of the conflict
- transportation modes tasked to transport personnel, munitions cargo, and nonmunitions cargo from FSLs to FOLs
- transportation modes available at each FSL at the end of a certain time period.

FSL Capability and Capacity

ROBOT selects the optimal FSL network from among a range of candidate FSLs, which are specified by the user. The user may input a wide selection of FSLs across a theater or a more limited set of options for comparison, depending on the scenarios to be analyzed. For each FSL, the user inputs the parking space, runway length and width, fueling capability, and capacity to load and off-load equipment. Runway length and width are key planning factors and are commonly used as first criteria in assessing whether an airfield can be selected.[23]

Afloat Prepositioning

ROBOT also allows the option of storing combat support resources (munitions and nonmunitions) aboard an Afloat Preposition Fleet (APF). Although afloat prepositioning does offer additional flexibility and reduced vulnerability versus land-based storage, the APF is much more expensive than land-based storage and presents a serious risk with regard to deployment time. Even if we assume a generous advance warning to allow for steaming toward a scenario's geographic region, it can be difficult to find a port that is capable of handling these large cargo

---

[23] ROBOT does not model factors such as hardness, altitude, and temperature range. In reality, planners must take these factors into account.

ships. The requirements placed on the port, including preemption of other cargo movement, also restrict the available ports an APF can use.

Airlift and Airfield Throughput Capacity

Timely delivery of combat support materiel is essential in any operation. However, a mere increase in the aircraft fleet size may not improve the deployment timelines. Planners must also consider the throughput capacity of an airfield. The MOG capability (i.e., the number of aircraft that can be simultaneously serviced on the ramp, whether refueling or loading/unloading cargo), for example, directly contributes to deployment time. In a scenario with heavy throughput demands, an airlift planner may posit that additional aircraft can accelerate the delivery of that cargo. However, if a key en route base lacks sufficient MOG, the ground capacity for servicing, refueling, and/or unloading aircraft—not the number of available aircraft—will prove to be the bottleneck to throughput.

ROBOT accounts for this by modeling the MOG for each FSL. FSL MOG constraints are defined in such a way as to account for both vehicle *space on the ground* and vehicle *ground time.* The MOG at each FSL is modeled separately for each class of vehicles, because air, ground, and sea vehicles are assumed to use different loading equipment. Each FSL is assumed to have a maximum number of vehicle spaces allowed for loading for each class at any one time. We assume that vehicles of different types and sizes consume different fractions of this loading space for different periods of time.[24] The MOG constraints similarly restrict FOLs based on the unload space available at each location.

Distance from FSLs to FOLs

Distance from FSLs to FOLs can impede operations. As the number of airlift aircraft increases, the difference in deployment time caused by distance becomes less pronounced. Adding more airlifters to the system will reduce the deployment time, albeit at a diminishing rate, until the deployment time levels off as a result of MOG constraints. ROBOT takes these factors into account when determining transportation requirements for a given basing posture.

Demand

A demand constraint requires the cumulative arrivals by a specific time to satisfy at least a pre-specified percentage of the cumulative demand. The model calculates the total time required to load a certain vehicle at a particular FSL, transit to an FOL, and then unload. FSL storage constraints limit the space available for munitions and nonmunitions. The model computes the demand requirement and assures that storage capacity is satisfied by

---

[24] ROBOT draws on prior RAND work to account for the factors that affect service times. See Stucker et al., 1998, and Stucker and Williams, 1999.

- selecting appropriate FSLs to meet the cumulative demand for each commodity at FOLs by the desired time
- optimally allocating the commodity sent from an FSL to an FOL via the appropriate mode of transport (subject to size limitations and restrictions concerning the types of materiel that can be carried)
- defining minimum units of storage needed for an economically feasible FSL at a given location
- computing additional square feet of storage space needed beyond the current level at a given FSL.

Cost

A main objective of the model is to reduce the total cost of supporting an operation while meeting the time-phased operational demand for combat support resources. ROBOT includes cost estimates for construction and/or expansion of facilities, operations and maintenance (O&M), and transportation for peacetime and training missions. Differences in regional cost-of-living or country cost factors are incorporated in each cost category.

By capturing each of these different categories of cost in the objective, ROBOT can capture various trade-offs in exploring FSL options. For example, introducing a new FSL into a theater may incur significant initial construction costs. However, those may be offset by reductions in transportation costs from an FSL more aptly positioned than existing sites. Moreover, the user has the option to exclude an FSL from the list of candidate sites. By running ROBOT with and without a specific FSL included in the solution set, the user can determine that location's cost benefit to the overall FSL network.

Anti-Access/Area Denial Considerations

The question of access deserves careful consideration and must be addressed before each conflict or operation. However, rather than eliminating some potential FSL sites *a priori* because of potential access problems, ROBOT first selects an optimal set of sites based on other factors (such as minimal cost or fastest closure times, as described above). The user then can "force" specific sites out of the solution set if there is a reason to believe that these sites present access issues. This approach shows the economic cost of restricting the solution to politically acceptable sites or to bases outside specific threat areas.

Moreover, to hedge against uncertainty in the future security environment, the user can test the robustness of an overseas combat support network across multiple series of possible scenarios. The model allows the user to perform what-if analyses by removing (or adding) locations or sites to the overall solution and observing the impact on overall operational performance.

Outputs

Based on the above inputs and constraints, the model finds the lowest-cost FSL network that satisfies the operational requirements over a given time horizon. That is, the costs of conducting training and deterrence missions are minimized, while the solution set is constrained to have the storage and throughput required to meet contingency scenarios should deterrence fail. The time-phased demands associated with these large contingencies ensure that the FSL network is capable of supporting large demands. Specifically, the formulation minimizes the net present value of opening and operating facilities, along with peacetime transportation costs, over a specific time horizon, to meet operational requirements. ROBOT outputs a transportation plan and reports the time needed for FOLs to achieve IOC and FOC. The model can also be used to determine FSL postures that meet other objectives, such as minimal deployment time or minimal number of airlifters required. The results of this analysis should yield global portfolios of FSL structures and combat support materiel allocations, including tables of metrics (such as allocation policy locations, technologies, and costs) that will allow policymakers to assess the merits of the various options.

Illustrative Model Runs

We illustrate ROBOT's capabilities using two scenarios: a relatively complex, real-world scenario based on prior PAF research and the simpler, notional scenario described in the appendix. The first demonstrates ROBOT's ability to optimize an FSL and transportation network from a range of possible FSLs to support multiple missions. The second shows how ROBOT compares the costs of supporting the Consolidated and Dispersed Beddowns in our simple notional scenario.

Two Simultaneous Contingencies in Southeast Asia

In the first scenario, we are interested in the minimal cost to meet the demands generated by two simultaneous contingencies in Southeast Asia: a training operation in Singapore and a humanitarian relief operation in East Timor. Demands were calculated using START.

For illustration purposes, we assume there are five potential land FSLs to choose from: Andersen Air Force Base (Guam), Clark Air Base (Philippines), Darwin (Australia), Paya Lebar (Singapore), and U-Tapao (Thailand). A sixth option is a munitions ship, *MUN2_Guam*, which is based in Guam. ROBOT will determine the optimal FSL network from among all six options. The user sets the MOG, land throughput, and seaport capability at each FSL.

ROBOT automatically determines the set of FOLs from the selected contingencies. In this case, Paya Lebar is opened to support training in Singapore, and U-Tapao is opened to support the humanitarian relief operation in East Timor. The user must define the time-phasing of the requirements by assigning values for the IOC (in days), the FOC that marks the end of operations, and the throughput for each FOL. The MOG and other parameters can also be adjusted. The model selects an optimum set of transport vehicles, but it also allows the user to assign transport vehicles, if needed. For this illustrative example, we assume that there are eight C-17s, 18 C-130s, two high-speed sealift (HSS) vessels, and 150 trucks available to deliver resources from FSLs to the FOLs.

As discussed above, the model may be used to minimize cost, deployment time, or number of airlifters required. In this example, we run the model to minimize cost. Based on the above inputs, the model identifies an optimal FSL and transportation network. The transportation cost, facility operating cost, and facility construction cost for the network are shown as the blue bars in Figure 3.2. The quantity of materiel (in tons) shipped from each FSL is displayed as the blue bars in Figure 3.3.

We ran an excursion on this scenario to show the cost impact of losing access to a candidate FSL. If we assume that either Thailand's government denies access to the materiel stored in U-Tapao or for other reasons the Air Force's leadership decides to bypass Thailand, the model can find alternative options for meeting the demand at a new cost (in terms of both time and funds). The results of this simulated denial of access are shown as the red bars in Figures 3.2 and 3.3. Without access to U-Tapao, system costs increase. The network requires additional investment to expand capacity at Paya Lebar and to open FSLs at Clark and Darwin. Network transportation costs also grow without access to facilities in Thailand.

Figure 3.2. Cost Comparison Between Baseline and Denied-Access Environment

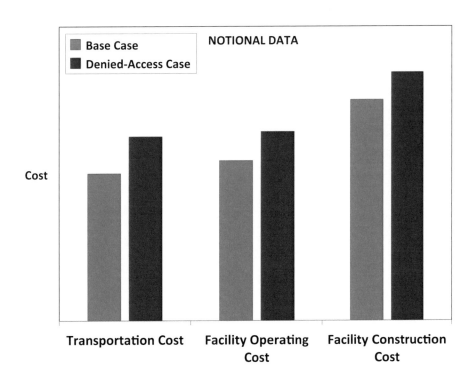

Figure 3.3. FSL Options for Base Case and Denied-Access Environment

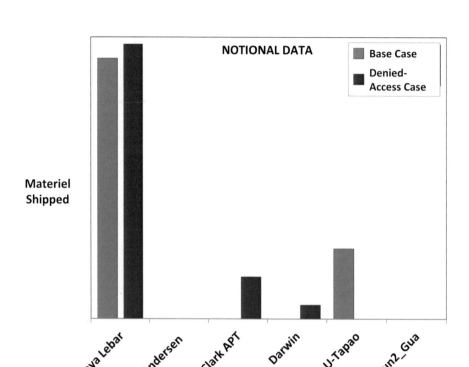

Notional Illustrative Scenario

A second scenario examines the implications of air base consolidation and dispersal. The appendix describes the properties of the Consolidated and the Dispersed Beddowns, and the results from Chapter Two provide the combat support requirement for each of the beddown models. We ran both beddowns using ROBOT with the objective of minimizing overall costs while meeting operational demands. The cost results are shown in Figure 3.4.

Figure 3.4. Cost Comparison Between the Consolidated and Dispersed Beddowns

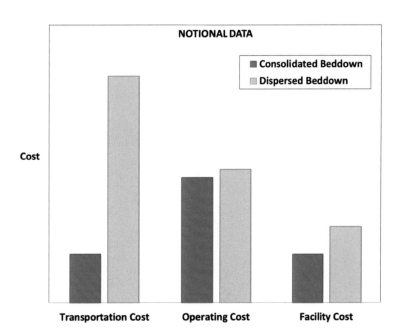

The impact on the transportation cost of operating from two additional bases is very clear, with cost increases in every category. In effect, Blue is forced to open, prepare, and stock two new bases, thereby increasing transportation cost and other operating and facility costs.

Conclusion

The illustrative runs presented in this chapter are limited to specific scenarios and theaters of operation. In actual model runs, ROBOT can evaluate the costs and risks of prepositioning postures to meet a range of scenarios on a regional or global scale. Prior PAF analyses suggest that a globally managed posture can support a robust set of contingencies at lower cost and with lower risk (e.g., from network disruptions) than a regionally managed posture.[25] In the CODE analysis, we can build on this prior work to include a range of A2AD scenarios and recommend low-cost prepositioning strategies as part of a global network.

---

[25] See Amouzegar et al., 2006.

4. Theater Air Base Vulnerability Assessment Model

Having identified the combat support resource and storage requirements to support air operations in a denied environment, we next consider air base vulnerability to threats and how threat mitigation options affect operational performance and cost. PAF developed TAB-VAM, a Monte Carlo simulation model, to analyze the complex trade-offs among basing strategies and mitigation options. The model allows the user to explore a wide range of scenarios, aircraft beddowns, base recovery capabilities, infrastructure investments, passive and active missile defense options, and CONOPS.

The major inputs and outputs to TAB-VAM are shown in Figure 4.1. Inputs include a database of air bases, a Base Manager File, a time-phased aircraft beddown, and an enemy campaign strategy. The primary output is the estimated sortie generation for each air base by aircraft type and by day of conflict. The model also displays the factors that affect sortie generation, which consist of enemy missiles fired, number of missiles intercepted, runway closure times, number of runway craters repaired, number of parked aircraft damaged, number of fuel tanks destroyed, tankers available for air-to-air refueling (AAR) purposes, amount of fuel demanded of these tankers by aerially refueled aircraft, and amount of fuel consumed.

Figure 4.1. Major TAB-VAM Inputs and Outputs

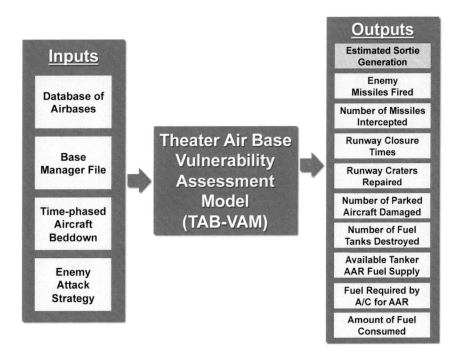

23

TAB-VAM is designed to assess the relative impact of various user inputs on sortie generation. For example, the user could choose to change the list of air bases; alter the aircraft beddown; invest in hardening, recovery, or redundant capabilities at some air bases; invest in active missile defenses; increase or decrease the enemy's capabilities; and/or modify the enemy's attack strategy. When combined with cost estimates for various investments, TAB-VAM can also be used for cost-benefit analyses. As discussed in Chapter Five, PAF developed a related modeling tool (TAB-ROM) that performs these analyses using TAB-VAM.

TAB-VAM is written in the programming language Java, which was selected for its speed and portability. Many of TAB-VAM's input files are in Extensible Markup Language (XML) format, with some data files instead being in Comma-Separated Value (CSV) format, and static information about air bases can be stored in a Microsoft Access database. Each run, defined as a user-specified number of Monte Carlo iterations for a particular set of input files, produces a single output text file with the outputs embedded as comma-separated tables. The user may choose a small number of iterations to keep the run time short or a large number to minimize the variance in results. A common number of iterations is 1,000, which results in minimal variance in results. It should be noted that, given reasonable inputs and computing resources, 1,000 Monte Carlo iterations can be completed in less than one minute.

Currently, there is no graphical user interface (GUI) for operating TAB-VAM. Instead, the user must modify all inputs or assumptions in the XML files with a text editor and then run TAB-VAM from a command prompt. The CSV components of the output file can be imported into Excel and the results analyzed and graphed as desired.

Inputs

This section details the major inputs to TAB-VAM, which include a database of air bases, a Base Manager File, an aircraft beddown, and an enemy attack strategy.

Database of Air Bases

The database contains the full set of possible air bases that are available for use in TAB-VAM, which is currently limited to bases in PACAF but could easily be extended to include any commercial or military air base. The database contains multiple tables of capabilities necessary for the logistical generation of aircraft sorties (e.g., runways, fuel storage). Associated with each capability are specific parameters that allow one to estimate a base's vulnerability to different enemy weapons. For example, for runways, the relevant parameters are number of runways, length of each, and width of each.

The focus of this database is on the specific capabilities that enable Blue to sustain sortie generation under attack. The database is not a comprehensive list of all facilities at an air base, nor does it include a full geographic layout of the airfield. For example, parking areas are treated as distinct from runways, though in reality, an attack on one might cause collateral damage on the other. The database does not provide enough information for a full damage assessment for an air base or include the information required for collateral damage assessments.

Base Manager File

The Base Manager File is an XML file that specifies the assumptions to be used in the current model run. Whereas the database of air bases includes every air base TAB-VAM can import into the simulation, only those bases specified in the Base Manager File will be imported as part of a user-specified beddown. Also, theater-wide defaults (e.g., fuel resupply rate, location of the fight, etc.) can be specified in this file. For further granularity, everything that can be specified at a theater-wide level can be specified at the individual base level, as can additional investments for those bases (e.g., shelters, ADR teams, missile defense assets, etc.).

Aircraft Beddown

TAB-VAM requires three input files for aircraft beddown information. The first is a CSV file that specifies the initial Blue beddown of aircraft. The second is a CSV file that handles the time-phasing of aircraft, which allows aircraft to be added, subtracted, or moved from each base at any point during the simulation. The third is a CSV file with physical characteristics of all the aircraft in the beddown. The inputs for the beddown files are

- base
- aircraft name (F-22, EA-6B, etc.)
- number of this aircraft at each base
- target daily sortie rate for this aircraft at each base
- aircraft type (fighter, tanker, etc.)
- aircraft size (large or small)[26]
- aircraft role (cruise missile defense [CMD], tanker, etc.)
- aircraft air-to-air refueling requirement (discussed below)
- fuel capacity
- maximum range per sortie
- missile loadout per sortie.

The user can generate and compare multiple beddown cases to test the implications of dispersal, concentration, and other force posturing strategies.

Enemy Attack Strategy

The enemy attack strategy is specified in a single XML file. It contains the enemy launch sites, missile inventory, and allocation scheme (discussed further below). Each launch site is a latitude and longitude used to determine the distance to each Blue air base. The inventory is the

---

[26] Aircraft size is used only to determine the minimum operating surface required for takeoff, as discussed further below. For this purpose within the model, it is sufficient to designate a given aircraft as either "large" or "small."

set of missiles to be used for attacks on Blue runways, aircraft parking, and fuel storage, excluding missiles set aside for other missions. For each type of missile, the following information is specified:

- name
- type (ballistic or cruise)
- minimum range
- maximum range
- number available
- circular error probable (CEP)[27]
- reliability (the inverse of failure rate)
- number of submunitions
- submunition dispersal radius for runway attacks
- submunition dispersal radius for parking attacks.

Model Algorithms

TAB-VAM is a Monte Carlo model that simulates the progression of a conflict, monitoring the effectiveness of Red's attacks and Blue's ability to generate sorties in the presence of those attacks. The conflict's duration is set by the user as a number of days. Typically, TAB-VAM steps through the conflict day-by-day, thus defining a "timestep" of 24 hours. The user has the option of choosing a shorter timestep. For example, should the user choose four timesteps per day, TAB-VAM will simulate the conflict in six-hour increments. As shown in Figure 4.2, each timestep includes four phases: Blue asset replenishment, enemy attack allocation, implementation of the enemy attack, and Blue damage assessment and recovery. Longer timesteps allow Red to levy larger attacks by allocating its quiver across fewer salvos. However, this can provide Blue a longer period between salvos for assessment and recovery. On the other hand, shorter timesteps can prove to be more disruptive to Blue's damage recovery efforts, but spreading Red's finite quiver across a higher frequency of attack diminishes the impact of each individual salvo.

The first phase, asset replenishment, handles the time-phasing of assets, replacement of destroyed assets, and regularly scheduled replenishment (e.g., daily fuel resupply). The enemy attack allocation phase determines which enemy missiles will be fired during the timestep and allocates them to attack particular Blue assets. In the enemy attack phase, each of the assigned attacks is carried out, including Blue's defense against those attacks. Finally, Blue attempts to determine what the damages were and to repair destroyed assets as possible. We describe the algorithms associated with each phase of a timestep below.

---

[27] CEP defines the distance within which 50 percent of the missiles are expected to hit. A small CEP indicates that a missile has high precision and can effectively hit a small target.

Figure 4.2. TAB-VAM Timestep

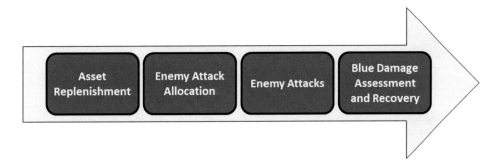

Asset Replenishment

The primary goal of asset replenishment is to allow Blue to add, move, and replace assets. Blue may choose to time-phase entire bases, aircraft, or missile defenses into the fight. It is important to remember that TAB-VAM is not a game; the analyst may not inspect the damage after each timestep and react to it. Rather the time-phasing of assets must be specified before the simulation runs occur. Outside of time-phasing, the model seeks to replace assets previously destroyed. For example, there is a user-specified parameter for the number of days it takes to replenish damaged aircraft. Also, asset replenishment includes regular replenishment capabilities, such as the user-specified daily intake of fuel for each base.

Enemy Attack Allocation

The second phase of each timestep allocates the enemy's missiles at particular Blue assets. TAB-VAM has two primary ways of handling this allocation: (1) automatic attack allocation and (2) manual attack allocation.

Automatic Allocation

In TAB-VAM's automatic asset allocation, illustrated in Figure 4.3, the user specifies the basic enemy *targeting strategy* (i.e., an allocation of missiles to bases), an *attack vector* (i.e., an allocation of missiles to specific assets on bases), and the percentage of missiles to be expended in the first volley. Based on these inputs, TAB-VAM allocates missiles to specific assets on specific bases at each timestep, using the algorithm described below.

Figure 4.3. Automatic Asset Allocation

The first step in the automatic allocation process, taken at the beginning of each Monte Carlo iteration, is to split the ballistic and cruise missiles among the different types of assets based on the user-specified percentages. This apportionment represents a single attack vector.

In this report, we allow ballistic missile targeting to vary between parked aircraft and runways at intervals of 25 percent (i.e., 100/0, 75/25, 50/50, 25/75, or 0/100). Similarly, we allow cruise missiles to vary between fuel and parked aircraft. This results in 25 attack vectors, as depicted on the grid in Figure 4.4. (We did not model the effects of collateral damage, though it is important to note that such effects might occur.) Different attack vectors will have different implications for U.S. and allied sortie generation, and would call for different combinations of mitigation measures. Later in this chapter, we will show theater-wide operational sortie generation for *one* of the 25 attack vectors, then the range of results *across the 25 attack vectors*. In Chapter Five, we will show how the optimization model, TAB-ROM, seeks the most robust mix of resources to improve sortie generation *across a user-defined portion of the tradespace* (e.g., the most effective Red attack vectors).

Figure 4.4. Tradespace of Attack Vectors Against Air Base Resources

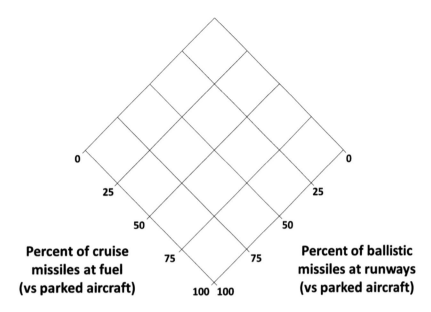

For each timestep, TAB-VAM determines the percentage of weapons that will be fired for each asset type. The initial volley specifies the percentage of weapons that will be fired in the first timestep of each Monte Carlo iteration, while all subsequent timesteps will have an equal allocation of the remaining missiles. For instance, for a scenario with one timestep per day and six days, if 50 percent of the missiles are fired on the initial volley, then each subsequent day will receive 10 percent of the total missiles. If the number of missiles does not divide evenly, additional missiles will be expended earlier in the simulation. In the above scenario, for example, if there were 13 missiles available, the allocation for days one through six would be six, two, two, one, one, and one missile(s), respectively.

TAB-VAM next allocates those missiles among the different bases according to the targeting strategy. The user selects one of three types of allocation: (1) TAB-VAM values the assets at each base equally, regardless of the number of assets, the beddown, etc.; (2) TAB-VAM uses the total number of aircraft available at a base to value the assets at that base; or (3) TAB-VAM uses the total number of a particular aircraft type (e.g., tankers) at a base to value the assets at that base. These options allow the user to explore the implications of different targeting strategies on Blue sortie generation, and thus to gain a comprehensive understanding of what the adversary might do.

Next, TAB-VAM divides the value of each base among the assets at that base. For runway assets, the value is split evenly among the different runway cutpoints. For parking assets, the value is split based on the size of parking aprons, and evenly among shelters. For fuel assets, the value is split evenly among fuel tanks.

Finally, TAB-VAM allocates missiles to individual assets, using the maximum marginal return assignment algorithm.[28] This algorithm allocates each missile based on the maximum marginal damage of missile:asset pairs. To determine the expected marginal damage of firing a single missile at a single asset, the value of the asset is multiplied by the probability that the missile will completely destroy the target asset. This probability includes the reliability, CEP, dispersal radius, and range of the missile (if the base is outside the range of the missile, the probability is set to 0). TAB-VAM will calculate this value for each missile:asset pair and select the pair with the maximum expected marginal damage to allocate a single missile at the asset. TAB-VAM will then reduce the number of missiles and decrease the value of the asset by the expected marginal damage of the pair. After the list of available missiles and the list of asset values have been updated, TAB-VAM will repeat the process until no missiles remain.

Manual Allocation

The manual attack allocation allows the user to specify in detail the enemy attack strategy. The user inputs a CSV file that includes all the missile attacks for a single iteration of the model, which is then repeated for each Monte Carlo iteration. This includes specifying for every timestep the number and type of missiles that will attack each type of asset (runways, parked aircraft, fuel system) at each base. For instance, two of "ballistic missile 1" might be sent to attack "runways" at "Base 1." TAB-VAM will still determine which individual assets to attack. For instance, given the previous example, if "Base 1" has a single runway, it will be attacked with both missiles, but if it has two runways, TAB-VAM will determine whether to fire one missile at each runway or both missiles at one runway, based on the logic in the automatic missile allocation. The manual allocation option is useful if one wishes to test a specific enemy attack strategy.

Battle Damage Assessment

One option specified in the Enemy Attack Strategy input file affects both the manual and automatic allocation schemes: the enemy's ability to perform battle damage assessment (BDA). Currently, TAB-VAM allows for two BDA settings: (1) no BDA or (2) perfect BDA. For the case of no BDA, the above allocation schemes do not change. For the case of perfect BDA, however, the enemy will not allocate missiles at targets currently destroyed. For instance, in the case of runway attacks, the perfect BDA case will not fire missiles at runway cutpoints not currently open, while the no-BDA case will simply fire at all cutpoints. Similarly, if the user has selected one of the asset valuation methods that includes the number of aircraft, the no-BDA option will use the total number of aircraft at the base, while perfect BDA will use only the number of undamaged aircraft.

---

[28] Kolitz, 1988.

Running Multiple Attack Vectors

While each model run reflects a single enemy attack vector (for a given targeting strategy), TAB-VAM makes its greatest contribution to force planning when we examine a range of attack vectors for a given scenario. This captures the inherent uncertainty of enemy attack plans and helps identify the most robust combination of Blue beddowns and base infrastructure options. In the CODE analysis, we run 25 attack vectors, which capture the percentage of cruise missiles fired at fuel storage versus parked aircraft and the percentage of ballistic missiles fired at runways versus parked aircraft. We illustrate the results of such an analysis using the notional scenario at the end of this chapter.

Having determined the enemy missile allocation, TAB-VAM next runs several distinct algorithms to model the effects of Blue missile defense; Red attacks on runways, aircraft parking, and fuel storage; and Blue options for mitigating damage. We discuss each below.

Missile Defense

After Red missiles have been allocated to certain targets and fired, they are subjected to Blue missile defenses. The available missile defense platforms in TAB-VAM are Terminal High Altitude Air Defense (THAAD) missiles, Patriot missiles, Aegis ships, and aircraft used in a missile defense role. The user completes three tables to define the performance of these assets: one that describes the performance of the interceptors, one that provides the interceptor loadout for each missile defense platform, and one that describes the defaults for each type of asset. The interceptor table lists, for each interceptor, the type of missile it defends against (ballistic or cruise) and the expected probability of kill of a single interceptor against a single incoming threat. The missile defense loadout table identifies, for each Monte Carlo iteration, the types of interceptors, the total number of launchers per interceptor type, and the total number of interceptors per launcher available for each missile defense platform. The missile defense asset table specifies, for each missile defense platform, the following:

- coverage area of the platform (single base, base cluster, or entire theater)
- domain of the platform (air, land, or sea)
- shot doctrine (shoot, shoot-shoot, or shoot-look-shoot).

The Base Manager File also specifies missile defense clusters (groups of bases that are treated as a single missile defense unit), and the missile defense assets are pooled together to defend against incoming attacks. Assuming that air-based missile defense platforms provide longer-range protection, we give these aircraft first priority in defending a base. Sea-based missile defense is the next priority in protecting a base against missiles not shot down by air-based missile defense. Land-based missile defense is the last set of platforms that can actively engage incoming missiles. If there is more than one missile defense asset for a given location and domain, the type of asset used is drawn randomly but weighted by the number of interceptors available to each asset for the raid. Currently, TAB-VAM does not support Red targeting of Blue missile defense assets, except for parked CMD aircraft, which may be targeted along with other parked aircraft.

TAB-VAM allows the user two options to augment missile defense at each cluster of bases. The first option is to add additional missile defense assets to a base. Inside the Base Manager File, the user can give each base an additional number of any of the four types of missile defense assets. The second option is to add additional interceptors to a base. While this will not allow the user to overcome per-raid maxima, it does overcome munitions shortages.

After missile defense has been executed, TAB-VAM determines the number of surviving missiles ("leakers") and models their effects on runways, parked aircraft, and fuel systems, as described below.

Runway Attacks and Mitigation Options

Each aircraft modeled in TAB-VAM has a minimum operating surface (MOS) necessary for takeoff and landing, depending on whether the aircraft is designated as large or small. In attacking Blue runways, Red's goal is to deny the MOS by attacking specific runway "cutpoints." For example, a 10,000-ft runway with a 3,000-ft-long MOS would require three cutpoints, creating four sections that would each be shorter than 3,000 ft.

Figure 4.5 shows the effects of a missile attacking a cutpoint. The dark gray area represents the runway. The missile is targeted at the center of the cutpoint, indicated by the circle with an X through it. First, CEP effects are applied to determine where the missile actually lands (indicated by the X). Given that landing point, each submunition is individually modeled (the small circles), falling somewhere within the dispersal radius of the missile landing point (the large circle). Submunitions that fall somewhere along the cutpoint become craters. Based on these effects, TAB-VAM determines whether or not a MOS is available and, if not, the minimum number of craters that must be repaired to restore the MOS. If any cutpoints can be opened before the end of the timestep, and if there are additional missiles allocated to runway attacks not allocated at particular cutpoints, then TAB-VAM will attempt to re-close the runways by re-attacking the base. Note that sometimes the attack MOS may be considerably smaller than the MOS required for aircraft to take off, and thus more than one open cutpoint may be required for the MOS to be deemed available.

Figure 4.5. Visualization of a Missile Hitting a Cutpoint

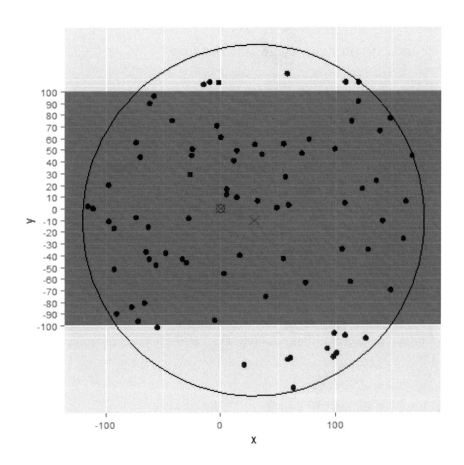

To mitigate the effects of runway attacks, Blue can employ two types of ADR: conventional Civil Engineering (CE) or "Advanced" ADR. The user chooses the amount of ADR (including zero repair capability) at each base as part of the Base Manager File. Conventional CE repairs are slow, but the equipment to effect repairs may be ready at hand. Advanced ADR, an improvement on conventional capability, includes better materials for repair of the damaged runway, but with similar damage assessment and EOD times. Four levels of capability have been modeled for Advanced ADR:

- small: 18 crater repairs per 10.5 hours
- medium: 54 crater repairs per 10.5 hours
- large: 90 crater repairs per 10.5 hours
- very large: 126 crater repairs per 10.5 hours.

TAB-VAM models the time required for runway damage assessment, EOD, preparation for the first crater repair, and subsequent crater repairs. After each attack, all runways are shut down to do both the runway damage assessment and EOD. Once those tasks have been completed, if there is no MOS available, the ADR teams will attempt to find and fix the easiest MOS. TAB-VAM will determine which MOS to attempt to repair first based on the beddown of each specific

33

base, because conventional CE teams repair the MOS differently for the two sizes of aircraft. If a MOS becomes available and a re-attack occurs, then runway damage assessment, EOD, and the prep time for the first crater will need to be repeated. If both small and large MOSs are available, then ADR teams will concentrate on extending the available MOSs as long as there are craters to repair.

Parked Aircraft Attacks and Mitigation Options

TAB-VAM models attacks on parked aircraft by determining which aircraft are on parking aprons, in shelters, and in the air. The user can specify either a theater-level default or individual-base level, or both, and then determine the percentage of aircraft that should be considered in the air at any time. The rest are then randomly assigned to shelters (as available), and the remainder are placed on parking aprons. When the attacks are launched, they first go through a missile defense check. Any leakers are then used to attack their individual targets. For parking aprons, the degree of damage is determined by the area of the apron that would be covered by the parking submunition dispersal radius. For shelters, TAB-VAM determines, based on the CEP of the incoming missile, whether the missile hits or misses the shelter. Any aircraft inside a directly hit shelter are damaged. For determining which parked aircraft are damaged, the algorithm does a random draw on the total percentage of parking apron area covered by parking attacks.[29] Any "hit" removes the aircraft from the fight.

To mitigate against parked aircraft attacks, the user can choose to add additional shelters to any given base. The Hardened Installation Protection for Persistent Operations (HIPPO) program has developed several designs for extremely durable aircraft shelters as well as options for less durable, but less expensive, shelters. These shelters range in size from a small shelter suited to protect a single small aircraft, to a medium shelter that protects three small aircraft, up to a shelter large enough to protect either six small aircraft or a bomber or tanker aircraft and the maintenance operation that supports it. A user may select one or a combination of these shelters at air bases in the scenario under examination. Although there are multiple options for each size of shelter in TAB-VAM, the illustrative results shown here will consider only a single type of shelter of each size for simplicity.

When TAB-VAM shelters aircraft, it takes into account only the size of the aircraft. TAB-VAM currently does not, then, take into account some valuation for the aircraft. For instance, an F-16 and an F-22 have an equal chance of getting put into a shelter, regardless of whether Blue would actually consider one of those aircraft more valuable. That said, TAB-VAM will prefer putting a single large aircraft in a large shelter over putting six small aircraft into the shelter.

---

[29] This is computed by dividing the attacked area (i.e., area over which incoming weapons' submunitions disperse) by the total area (i.e., the area of the ramp under attack). This is the percentage chance that an aircraft on that apron will be hit. Such a calculation is accurate in typical cases, where the aircraft are much smaller than the total area under attack.

At the end of the run, TAB-VAM calculates the number of parked aircraft damaged and factors this output into the estimated sortie generation.

Fuel System Attacks and Mitigation Options

There are currently two ways for the adversary to disrupt the fuel supplied to aircraft: attacks on a base's fuel system or attacks that hinder AAR using tankers. Each type of attack can have a substantial impact, as the destruction of on-base fuel resources will affect all sorties at that base, whereas hindering the tankers' ability to provide AAR could cause problems throughout the theater.

TAB-VAM models a base's fuel system by incorporating the fuel tanks used to store the fuel, the daily resupply of fuel, and the consumption of fuel by aircraft. Attacks on the fuel system, then, are focused on the fuel storage. Red missiles will go through a missile defense check, and then leakers will have CEP effects applied, and TAB-VAM will determine if the missile successfully hit the storage tank. In the Base Manager File, the user specifies the amount of fuel lost when a tank is hit. Also in the Base Manager File, the user may specify the amount of fuel that will be resupplied each day—as a percentage of the total fuel capacity of the base—and that fuel will be stored into any available space inside fuel tanks. When determining sortie generation, TAB-VAM will attempt to find fuel in tanks to fuel sorties, based on the "fuel per sortie" consumption rates specified in the Aircraft Beddown. At the end of the run, TAB-VAM reports the number of fuel tanks destroyed and the amount of fuel consumed, and factors these results into the estimated sortie generation.

TAB-VAM models AAR by setting up tanker orbits, calculating how much AAR is required for each aircraft in the beddown, and determining which aircraft can be supplied with enough fuel to complete their missions. As discussed earlier in this chapter, the TAB-VAM user enters the location of the fight, the maximum range of each aircraft, and which aircraft require AAR. For each aircraft that will use AAR, TAB-VAM calculates the optimal location for a tanker asset based on the location of the base and the aircraft's route to the fight. It then assigns tankers based on the combined optimal locations such that refueling can be completed for as many aircraft as possible using as few orbits as possible. After the orbits are placed, TAB-VAM allows aircraft to take off for the sorties the tanker orbits can support and keeps track of how much fuel gets consumed in the process. The following are important assumptions in this process:

- Aircraft that are not in a missile defense role fly to the fight and use 100 percent of their maximum fuel each time they travel 100 percent of their maximum range.
- Aircraft that are in a missile defense role do not fly to the fight. Rather, the time they are flying in their missile defense role is what is used to calculate tanker demand.
- Aircraft that are in a missile defense role require half their maximum fuel for every hour they are airborne.

- Unlike aircraft that fly to the fight, missile defense aircraft can perform their mission for a limited time without tanker support, but they must return to base to be refueled, resulting in lower probability they can detect and engage incoming missiles.

There are two methods of mitigating fuel system attacks in TAB-VAM. The first, additional fuel tanks, not only creates additional targets for the enemy, but also adds the volume of the additional tanks to the fuel available at the base. The second, fuel bladders, also adds fuel to the base and targets for the enemy. The main difference in TAB-VAM between adding bladders and adding fuel tanks is that bladders are a different size, which causes them to hold less fuel and have a different targeting footprint. Cost also differs between the two, as will be shown in Chapter Five. We assume that, because of strategic warning, all fuel assets on the base are full when the conflict begins. The user may constrain the number of bladders that can be employed, due to space limitations at a given base, as we do in the optimization analysis discussed in Chapter Five.

Unlike fuel system attacks, AAR can be disrupted by any attack that keeps tankers on the ground. As a result, investments in runway repair, aircraft shelters, fuel tanks and bladders, and active missile defense could all be used as potential mitigation strategies depending on the enemy attack vector.

Mitigation Using CEP Adjustment

If Blue were to have the capability to degrade Red missile performance by causing an increase to the missile CEP, the effect can be incorporated in TAB-VAM. In the Base Manager File, the user enters a CEP adjustment factor for ballistic and cruise missiles and specifies which bases, if any, have the CEP adjustment capability. For the missiles allocated to bases without a CEP adjustment, each missile will use the CEP specified in the missile inventory. For missiles allocated to bases with a CEP adjustment, the CEP from the missile inventory is multiplied by the cruise or ballistic adjustment factor, depending on the type of missile. Given the adjustment factor is greater than one, the result is a lower probability that the missile will hit its target (runway, parking area, aircraft shelter, or fuel tank). Since collateral damage has not yet been incorporated into TAB-VAM, increased CEPs typically result in increased sortie generation since the attacks are less effective.[30]

Adaptation

It is important to remember that TAB-VAM is not a game; it does not adapt either attack or defense strategies throughout the simulation. TAB-VAM requires that all operational assumptions be chosen at the beginning of the simulation run. If the enemy weapon allocation process provides too few weapons per day to close an air base's runways, the simulation will not

---

[30] In reality, less-accurate missiles could miss the intended target but hit a nearby target of the same type (e.g., other fuel storage on a fuel farm) or of a different type (e.g., aircraft parked near a targeted runway).

increase allocation on the next day or the following day. Similarly, if the allied beddown results in extremely limited sortie generation of a particular kind (e.g., tankers, defensive counter-air), then TAB-VAM algorithms have no ability to move those aircraft to a base that could generate more sorties. In reality, both sides would make adjustments during the conflict. The TAB-VAM user needs to review the outcomes of a TAB-VAM run and appropriately adjust the inputs to be as realistic and reasonable as possible.

Illustrative Model Run

The remainder of this chapter illustrates TAB-VAM outputs using the notional scenario described in the appendix. We begin by showing baseline results for the Consolidated Beddown. Next, we compare results for the Consolidated and Dispersed Beddowns, using average theater-wide sortie generation over the course of the scenario as the primary metric. Finally, we show how investing in additional mitigation options at specific bases alters overall performance. The purpose of this section is to illustrate the kinds of outputs TAB-VAM can generate rather than to draw substantive conclusions about the results of this highly notional scenario.

Baseline Results for Consolidated Beddown

The Consolidated Beddown consists of five bases on a series of island chains at various distances from the adversary mainland. We ran TAB-VAM using the baseline assumptions listed in the appendix, which include the Red quiver; Blue beddown of fighters, bombers, and tankers; Blue missile defense and base infrastructure; and a limited amount of Blue damage repair capabilities. For this baseline run, we assumed a Red targeting strategy that attacks all bases in proportion to the number of aircraft, and a single Red attack vector that fires 50 percent of ballistic missiles at runways, 50 percent of ballistic missiles at parked aircraft, 50 percent of cruise missiles at parked aircraft, and 50 percent of cruise missiles at fuel storage (we show the effects of multiple attack vectors in the next section).

The following sections show detailed results for Red missile allocation, missile defense intercepts, runway closure times, number of runway craters repaired, number of each type of parked aircraft damaged, number of fuel tanks destroyed, and the percentage of sorties generated for each type of aircraft at each base over the length of the conflict.

Enemy Missiles Fired

TAB-VAM automatically allocates missiles to specific targets at specific bases using the allocation algorithm described above. Figure 4.6 shows the results for our baseline case. Because Red missiles are allocated according to the number of aircraft at each base, Base1 receives more missiles than the other bases. Also, since Base 1 is the only base inside the range of short-range ballistic missiles (SRBMs), all the SRBMs are allocated to it (with 50 percent targeted at parked aircraft and 50 percent at runways). The MRBMs are divided between Bases 2, 3, and 4, and all the IRBMs are targeted at Base 5. Both the ground-launched cruise missiles (GLCMs) and the air-launched cruise missiles (ALCMs) have wider bands (though the GLCMs are not able to hit Base 5), so they are distributed according to value among the bases. It is notable that only fuel storage is attacked at Base 3. This is because Base 3, an allied base, has the smallest number of aircraft and is therefore considered to be a less valuable target. The next set of figures show how many missiles penetrate Blue missile defense and the effects of these survivors on runways, parking, and fuel.

Figure 4.6. Enemy Missiles Fired

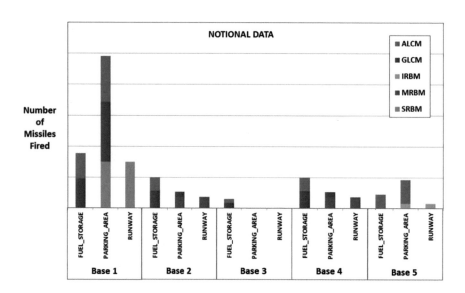

Missiles Intercepted

As discussed in the appendix, we assume that there is one Blue fighter squadron at Base 2 to perform CMD. Bases 2, 3, and 4 are treated as a single missile defense cluster; therefore, the squadron at Base 2 will engage enemy cruise missiles fired at any of these three bases, giving equal priority to each enemy missile fired at that cluster. As shown in Figure 4.7, the number of missiles intercepted is much higher on the first day because Red launches 50 percent of its missiles in the initial volley. All the missiles that survive Blue missile defense go on to strike their allocated targets, with the effects detailed below.

Figure 4.7. Missiles Intercepted by CMD Aircraft

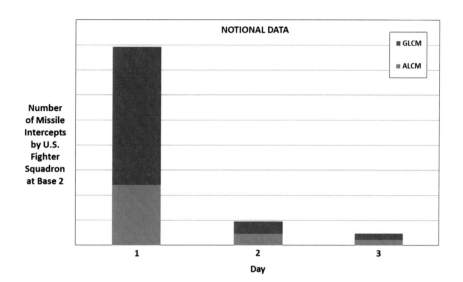

Runway Closure Times

Figure 4.8 shows the number of minutes per day that runways are open for small and large aircraft. Although not visible in the figure, Bases 3 and 5 are open 100 percent of the time, but for different reasons. Base 5 runways are struck by IRBMs, but the small Advanced ADR team is able to keep pace with repairs. Base 3 receives no missiles at its runways, as it is the least valuable target in MRBM range (as shown in Figure 4.6). Bases 2 and 4, however, are high-value targets, with many aircraft, only one runway each, and conventional CE. Therefore, Bases 2 and 4 are associated with significant time getting large aircraft off the ground (small aircraft fare better). Finally, Base 1 runways receive a large number of missile strikes, but the small Advanced ADR team is able to fix the runways relatively quickly.

Figure 4.8. Runway Availability

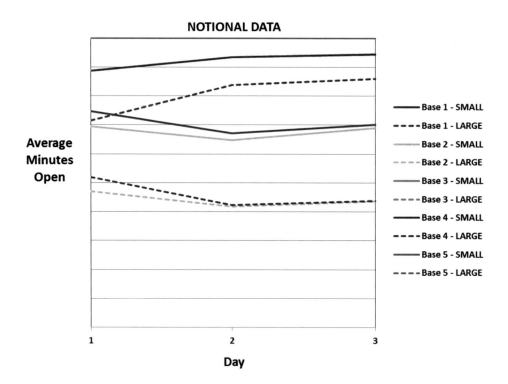

Runway Craters Repaired

The advantage of Advanced ADR is even more apparent when we examine the number of craters repaired at each base over the entire conflict, as in Figure 4.9. The conventional CE teams at Bases 2 and 4 are able to fix only a minimal number of craters, but Bases 1 and 5 are able to repair many more craters (recall that Base 3 suffers no runway attacks). These results are also useful for understanding the consumables (e.g., quickly curing concrete) required by Advanced ADR teams: We assume that Base 1's Advanced ADR team has unlimited consumables, which likely would need to be prepositioned to be available.

40

Figure 4.9. Runway Craters Repaired

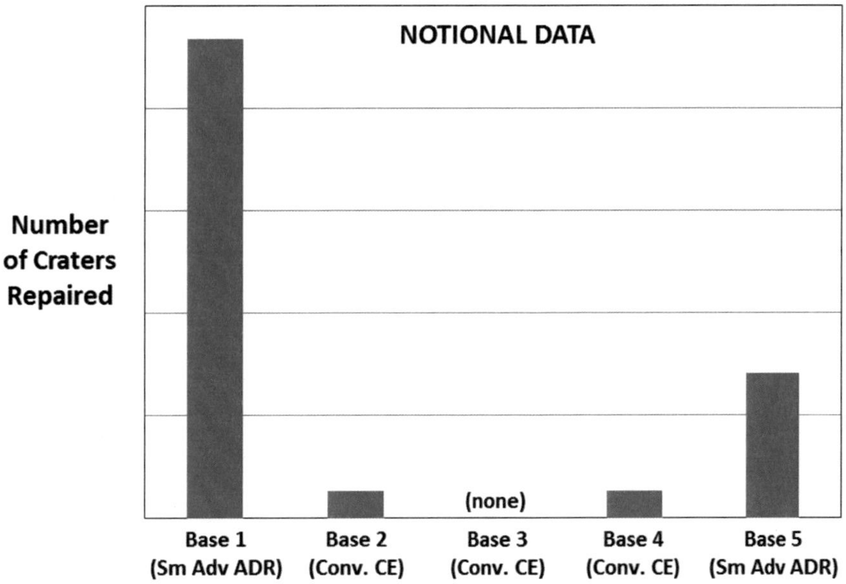

NOTIONAL DATA
Number of Craters Repaired
(none)
Base 1 (Sm Adv ADR)
Base 2 (Conv. CE)
Base 3 (Conv. CE)
Base 4 (Conv. CE)
Base 5 (Sm Adv ADR)

Parked Aircraft Damaged

We turn next to the number of aircraft damaged by Red's attacks on parking areas. Figure 4.10 shows the average total aircraft damaged at each base. Not surprisingly, damaged aircraft are greatest at Base 1, which is closest to the threat. Base 3 has no damaged aircraft because it suffers attacks against fuel storage only.

Figure 4.10. Parked Aircraft Damaged

NOTIONAL DATA
USfighter
MedTanker
LargeTanker
Bomber
Average Total Aircraft Destroyed
(none)
Base 1
Base 2
Base 3
Base 4
Base 5

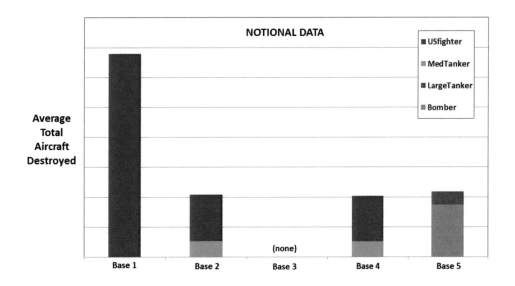

Fuel Tanks Destroyed

Another major factor in calculating sortie generation is the number of fuel tanks destroyed and the effect on fuel consumption. Figure 4.11 shows the average number of fuel tanks destroyed (across all Monte Carlo iterations) at each base. Although not shown, TAB-VAM also calculates the volume of fuel in those tanks and the total consumption. This allows the user to determine the amount of fuel susceptible to attack and thus where additional fuel tanks or bladders could prove advantageous.

Figure 4.11. Fuel Tanks Destroyed

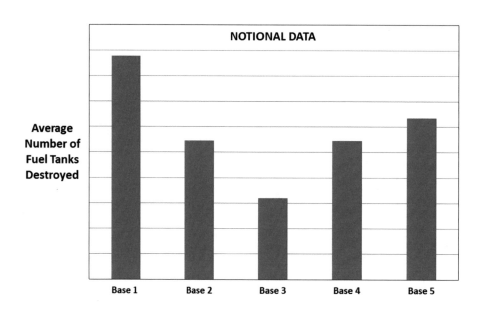

AAR Supply and Demand

The final piece of sortie generation captured by TAB-VAM is whether or not aerially refueled aircraft have adequate tanker support to complete their missions. Figure 4.12 shows the amount of fuel each base needs for AAR, the amount of fuel each base can supply for AAR, and theater-wide supply and demand totals. Supply totals are a function of how many tanker sorties are at a base. Demand totals depend on more factors, such as distance from the fight and role of the aircraft.

Base 1 needs very little AAR, since it is close to the fight. Base 2 also needs very little, since its missile defense fighters do not fly to the fight. Base 3 requires no AAR, since it is an allied base and all those aircraft stay local. Base 4 aircraft require more AAR than the first three bases since its fighter aircraft fly to the fight. Base 5 easily has the greatest AAR demand, since it is farthest from the fight and its bombers burn fuel at a greater rate than the fighters at the other bases.

In this notional case, there is a clear shortfall in AAR supply. It could be due to effective attacks by the adversary, not having enough tankers in the theater to begin with, or a combination of the two. In either event, this shortfall should cause bombers at Base 5 to get less than half their sorties off the ground, as shown in the next section.

Figure 4.12. AAR Supply and Demand

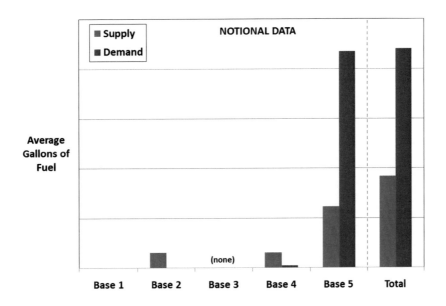

Primary Performance Metric: Sortie Generation

Having simulated Red missile attacks and their effects on runways, parked aircraft, and fuel, TAB-VAM calculates the percentage of planned sorties generated. This is the primary performance metric for any given model run and is used to compare different analytic cases (as discussed in the next section).

There are two separate outputs in TAB-VAM for sortie generation. The first, referred to as "Base" sorties, simply measures how many sorties can take off from each base, given the expected damages to runways, parked aircraft, and fuel storage. This measure of sortie generation does not take into account whether or not there is AAR available for the sorties to perform their missions. The second output, referred to as "Theater" sorties, does take AAR into account. Having both sortie generation outputs helps users determine when a sortie generation shortfall is related to AAR.

Figure 4.13 shows the Base and Theater sortie generation for each type of aircraft, at each base, on each day during the baseline simulation. Not surprisingly, the fighter sorties at Base 1 suffer most from close-range attacks on runways, parking, and fuel. Bases 2 and 4 perform somewhat better, with fighters doing better than tankers at each base. Base 5 does still better, being farthest from the adversary. The allied Base 3 performs best of all, mostly because it makes the least attractive target when Red attacks are allocated according to the number of aircraft.

As expected, the Base sortie result is always greater than or equal to the corresponding Theater sortie result since including AAR in the computation will never improve sortie generation. Some aircraft, namely the tankers and missile defense fighters, do not require AAR to generate sorties,[31] which causes the Base and Theater sorties to be equal for those aircraft. The rest of the aircraft have lower Theater sortie percentages since not all fighters and bombers have adequate AAR to get to and from the fight. The bomber sorties at Base 5 were particularly affected by AAR constraints, which is consistent with the previous section. If these AAR shortfalls did not exist, the two charts would have been identical. Since Theater sorties are more operationally relevant than Base sorties, the remainder of this report will use Theater sorties as the sole sortie generation metric.

Figure 4.13. Sortie Generation

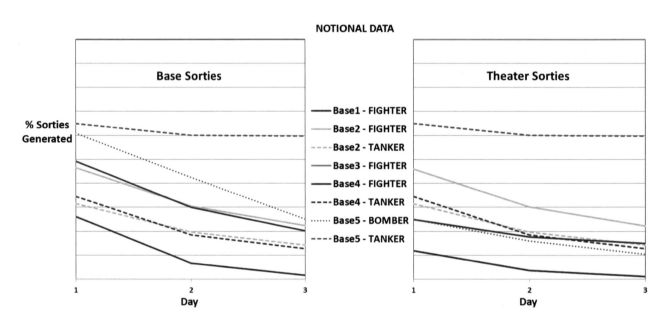

Comparing Model Runs

Having shown the baseline model run in detail, we next illustrate the types of comparisons that can be made using TAB-VAM results for various enemy attack vectors, beddowns, and investments in base infrastructure and damage repair capabilities. For simplicity, we assume a Red targeting strategy that attacks all air bases in proportion to the number of aircraft at each base.[32]

---

[31] Recall that missile defense fighters can take off if tankers are unavailable, but their effectiveness is lower.

[32] In the actual CODE analysis, we explore and compare a variety of targeting strategies.

Varying Enemy Attack Vectors

The baseline model run described above represents Blue performance against a *single* Red attack vector. Thus, it can be expressed as a single number: the average percentage of sorties generated theater-wide over the duration of the conflict.[33] However, as noted above, TAB-VAM produces the most robust analysis when one examines how a given beddown and base infrastructure case performs against a *range* of enemy attack vectors (within a given targeting strategy).

Figure 4.14 shows how the baseline Consolidated Beddown performs against a range of enemy attack vectors. The x-axis represents ballistic missile allocation, which varies between runways and parked aircraft. The y-axis represents cruise missile allocation, which varies between fuel storage and parked aircraft.[34] Since the sortie percentages shown in the contour represent Theater sorties, tanker dependencies are factored into the bomber and fighter sortie percentages. Thus, to ensure that the impact of tankers is not accounted for twice, we do not include tanker sorties in the contour (and for the rest of this report).

The baseline run described in the previous section appears in the center of the performance surface, which is significantly higher than some other parts of the surface. Thus, it represents an optimistic view of Blue capability. The steep drop-off on the front of the surface (facing the viewer) suggests that Red's best strategy is to fire 100 percent of cruise missiles at fuel storage (as opposed to parked aircraft) and to fire ballistic missiles at some combination of runways and parked aircraft (with runways being the best target). Thus, the goal for analysts is to find alternative beddowns and/or investments that will improve Blue performance across the entire surface (i.e., "lift the corners").

---

[33] The user may choose which types of aircraft sorties to include in this performance metric.

[34] This chart shows the trade-off between two assets for each type of missile. Currently, in TAB-VAM there are three possible assets that each type of missile could fire at. However, two modes are not shown on the plot, namely that ballistic missiles could be fired at fuel, and cruise missiles could be fired at runways. Although the data become more difficult to visualize as the number of possible permutations increases, TAB-VAM is able to determine the value of those weapon targeting trade-offs.

45

Figure 4.14. Average Sortie Generation Against All Enemy Attack Vectors (Consolidated Beddown)

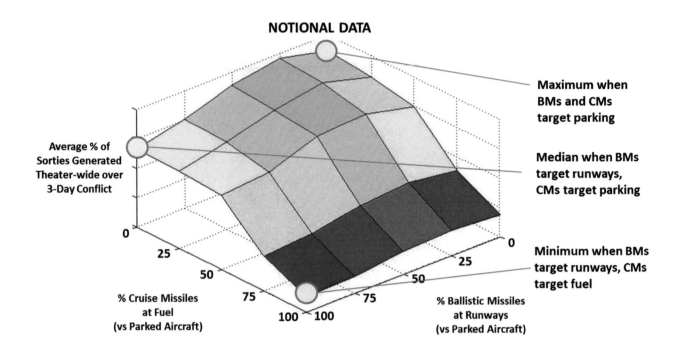

Comparing Beddowns

While the three-dimensional image in Figure 4.14 is useful for examining a single analytic case, it is difficult to visualize comparisons between different cases. A simpler way to depict results is to find the minimum, median,[35] and maximum sortie percentages from the 25 data points on the performance surface, as these points capture the best case for Red, the best case for Blue, and a point between the two. In subsequent figures, we represent this range as a "box plot" that shows minimum, median, and maximum values for a given model run.

A critical issue is how different beddowns compare in terms of sortie generation. Figure 4.15 compares average sortie generation for the Consolidated and Dispersed Beddowns. The additional bases in the Dispersed Beddown have only a few fuel tanks, so both performance surfaces have the same steep drop-off for fuel storage attacks. Consequently, the minimum and median points are similar. The maximum points for the two beddowns occur when parking is the primary target. The additional parking areas in the Dispersed Beddown complicate the adversary targeting, and therefore increase the sortie percentage.

---

[35] Median is the "middle" value in a list that is ordered from least to greatest. The median is used here instead of average because it is easier to pick out on the performance surface.

Figure 4.15. Comparing Beddowns

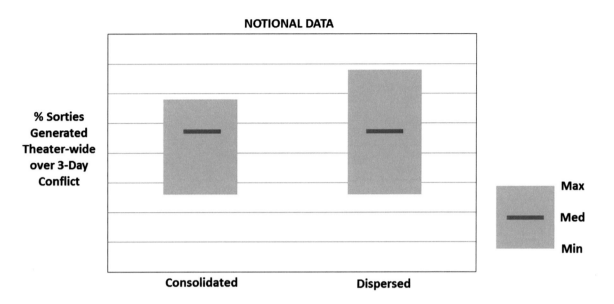

Comparing Investments in Mitigation Strategies

TAB-VAM can also help examine the relative effectiveness of different investments in base infrastructure and damage repair capabilities. We begin with a simple comparison, shown in Figure 4.16. The first box is the baseline performance for the Consolidated Beddown, shown in the previous figure. The second box shows the effect of adding to Base 4 one Fuels Operational Readiness Capability Equipment (FORCE) kit, which includes 24 fuel bladders and is discussed in Chapter Five. As expected, the augmented capability improves the minimum sortie point, since that point was lower before as a result of vulnerability to fuel attacks. However, it is important to observe that the median and maximum sortie generation are unaffected, since adding fuel bladders alone does not help against all types of attack.

47

Figure 4.16. Assessing Investment in Fuel Bladders

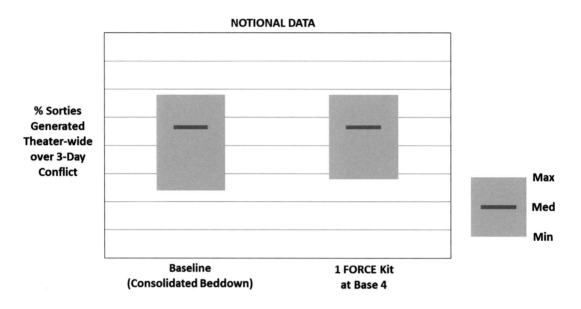

Next, we compare a variety of single-resource investments. For simplicity, we vary investments only at Base 4 (but the results reflect average sortie generation across the entire theater). The boxes in Figure 4.17 are ordered from least median sortie generation to greatest. Very few of the investment options affect the minimum sortie point, since most investment options do not directly address the vulnerability from cruise missiles attacking fuel storage. The median sortie point does not change much either, since most investment options shift only a portion of the performance surface, with a small overall impact on results. The best single investment overall (for this notional case) is to add a THAAD unit, as it has strong performance against both runway attacks and parked aircraft attacks.

Figure 4.17. Comparing Investments in Single Mitigation Types

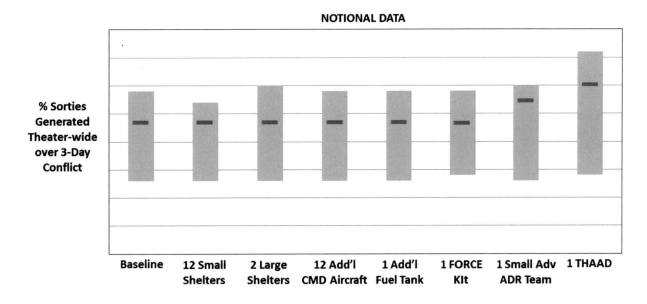

A THAAD unit may be more expensive than an Advanced ADR team and 12 small shelters, or an Advanced ADR team and a FORCE kit. TAB-VAM can illuminate trade-offs between different combinations of assets, as illustrated in Figure 4.18. For this notional case, the combination of one small Advanced ADR team and one FORCE kit performs better than a single THAAD unit in terms of minimum sortie generation.

Figure 4.18. Comparing Investments in Multiple Mitigation Types

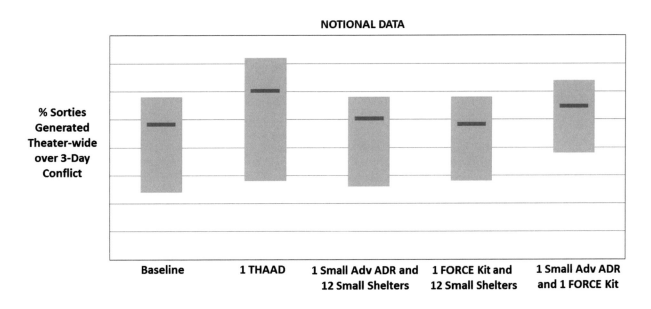

Another issue that TAB-VAM can help examine is how the placement of assets at different bases affects performance. The first box in Figure 4.19 represents the baseline. The second box represents a mix of investments in Advanced ADR, shelters, fuel bladders, and THAAD, all at Base 4. The third box shows the same mix of investments, but distributed around the theater. In this notional scenario, the dispersal of resources worsens overall sortie generation compared to the concentration of resources at one base. This is no surprise since Figures 4.17 and 4.18 show that single investments at bases have little impact on the performance surface, whereas combining investments at a base can have a more substantial impact. A more interesting result occurs when there are multiple investments for each base, which can be done in many different ways and will be explored in Chapter Five.

Figure 4.19. Assessing Asset Placement

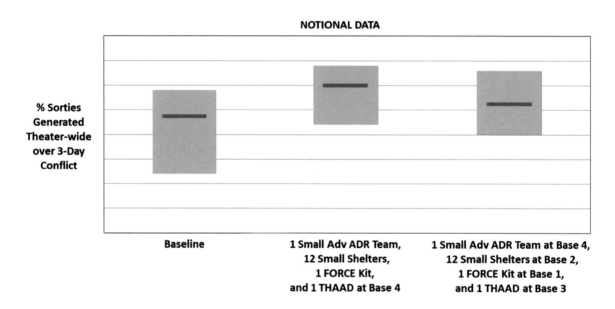

Of course, investments in base infrastructure and damage repair capabilities come at a price. Finding the optimal mix of resources within a given budget is a serious analytic challenge involving a massive number of permutations. It is precisely because of this combinatorial expanse that we developed TAB-ROM, an optimization model that uses TAB-VAM to select the most effective mix of investments at a given budget level (as discussed in Chapter Five).

Conclusion

TAB-VAM is a powerful tool for understanding the challenges and complexities of supporting air operations in denied environments. The model provides a robust, detailed, multilayered representation of the factors that can affect Blue performance, here measured as the ability to generate sorties over the course of a conflict. By allowing the user to vary many dimensions of Blue and Red operations and resources, TAB-VAM provides an assessment of uncertainties and risks. It also allows for a robust exploration of trade-offs between different basing strategies and resource investments. TAB-VAM can be expanded to examine a wider range of resources and investment options. When combined with cost estimates for the various mitigation options, it allows force planners to analyze various mixes of investments and determine an appropriate budget threshold to achieve desired sortie generation goals. In the next chapter, we examine how TAB-ROM is used to identify optimal investments at various budget levels.

5. Theater Air Base Resiliency Optimization Model

As discussed in Chapter Four, TAB-VAM allows the user to explore the effect of different air base facilities, aircraft beddowns, and enemy missile threats on sortie generation in denied environments. PAF developed TAB-ROM as a companion model that can determine the optimum investments in threat mitigation resources, such as hardened aircraft shelters, ADR capabilities, fuel storage redundancy and dispersal, and missile defense. TAB-ROM's strategy is to invest in the set of resources that maximizes sortie generation theater-wide, averaged over a range of adversary attack vectors (within a given Red targeting strategy). TAB-ROM specifies the quantity of each resource to be purchased, limits those purchases to a user-defined budget, and indicates where each resource should be positioned.

This chapter describes TAB-ROM's major inputs, algorithm, and outputs. The chapter ends with a notional illustration using the scenario described in the appendix.

Inputs

In this section, we enumerate the four key inputs to TAB-ROM: procurement options and cost data, the size of the investment budget, system constraints, and an objective function to guide the optimization.

Procurement Options and Costs

TAB-ROM requires the user to input a menu of specific investment options in missile defense, ADR, hardened aircraft shelters, and fuel storage, as well as the cost of each option. Table 5.1 summarizes the selections available to TAB-ROM in each category of passive defense. Note that the costs listed are notional.[36] An authoritative composition of Advanced ADR teams has not yet been finalized, and decisions regarding the layout of hardened aircraft shelters are currently provisional. However, the values shown are roughly consistent with current design estimates. Costs for fuel infrastructure, such as above-ground storage tanks and the bladders and pumps comprising FORCE kits, are much better defined. The values shown in the table for these fuel systems are cost estimates for the equipment needed to handle approximately 1.2 million gallons of fuel (the volume for each static, above-ground tank). A FORCE kit includes twenty-four 50,000-gallon bladders, or enough volume to hold fuel for one tank.

---

[36] These notional figures represent procurement costs only; they do not include sustainment or maintenance costs. Note that the actual CODE analysis uses actual cost estimates.

Table 5.1. Mitigation Options Available for TAB-ROM Portfolios

Capability Type	Option	Notional Cost ($M)
ADR	Conventional CE	0
	Small Advanced ADR Team	15
	Medium Advanced ADR Team	30
	Large Advanced ADR Team	45
	Very Large Advanced ADR Team	60
Aircraft Shelters	Small	20
	Medium	80
	Large	160
Fuel Storage	FORCE kits	2
	Static fuel tank	10

Investment Budget

Another important input to TAB-ROM is the size of the budget that the optimization will be allowed to allocate between the various investments. With a sufficiently large budget, TAB-ROM will be able to consider the sortie generation value of more expensive mitigation capabilities, such as hardened shelters and missile defense systems. However, a small budget may reveal very cost-effective, asymmetric damage mitigation investments that could assist in a relatively rapid recovery from damage dealt by the adversary's quiver of expensive weapon systems.

System Constraints

The user has the option of providing any constraints to the scope of investment portfolios TAB-ROM will be allowed to explore. The first key limiting factor involves restricted availability of Air Force UTCs. These shortfalls can become apparent in highly dispersed beddowns, where the user may run START (see Chapter Two) and learn that not all bases can be supported using current manpower and equipment UTCs. For example, let us assume that START reveals a potential deficit in manpower for fuel system UTCs. This, in turn, would suggest constraining the number of FORCE kits that TAB-ROM could procure.

Availability of base infrastructure forms a second major category of constraints. Often during site surveys and beddown planning processes, the user may learn that there are limited opportunities for constructing new infrastructure at a base. Small bases and locations near dense urban regions, in particular, may lack sufficient footprint to accommodate new large structures such as static fuel tanks or hardened aircraft shelters. Similarly, a site survey may determine that existing fuel storage tanks at an older airfield have degraded and require replacement. This may lead to a constrained floor for the number of new tanks that must be constructed at this location.

53

The user can incorporate any such known infrastructure bounds into TAB-ROM's procurement options.

A third category of constraints is the broader set of user-defined limiters. Constraints here typically allow for the user to shape TAB-ROM's solution to adhere to additional planning guidance that falls outside the realm of the first two constraint types. For example, the user could choose to tailor the placement of Advanced ADR teams only at bases within the inner missile threat ring or to procure new hardened fighter shelters only at FOLs. TAB-ROM is sufficiently flexible to accommodate most user-defined constraints.

The Objective Function

Finally, TAB-ROM requires the user to select the metric that will guide the selection of elements within the resiliency investment portfolio. By default, TAB-ROM will seek out the set of investments that maximizes the sortie generation of operational aircraft (i.e., fighters, bombers, SOF, and C2ISR)[37] across the entire theater, as averaged across a range of user-defined attack vectors (for a given Red targeting strategy).

Given the central role the objective function plays in an optimization, it is important to reflect on nuances in the three key elements touched on above. To reiterate, the default objective is to maximize all operational sorties, theater-wide, across a range of attack vectors.

By maximizing all operational sorties, the objective function does not distinguish between aircraft types. A bomber sortie weighs just as heavily in the objective as a fighter, SOF, or C2ISR sortie. Note that TAB-ROM, much like TAB-VAM, does not assess the combat effectiveness of sorties. TAB-ROM simply computes the sortie generation capability of investments in damage prevention and mitigation options. We address these weights further below, but by default, TAB-ROM treats all operational sorties as equal and seeks to maximize their total in the aggregate.

By maximizing all operational sorties theater-wide, the objective function does not take into account an aircraft's distance from the fight. A bomber's long-range strike mission will be as important as a close-in fighter aircraft's defensive counter air sortie in the objective function. Again, TAB-ROM does not evaluate these sorties' combat effectiveness or relative importance. Rather, it seeks to protect as many sorties as it can with investment opportunities that either protect assets or help the base recover from damage.

In considering the importance of one sortie type versus another, TAB-ROM interacts with TAB-VAM to offer a variety of ways to sort and present results. Thus, the user can choose to show sortie data for only certain types of aircraft. As mentioned in Chapter Four, we typically only show sortie results for bombers, fighters, C2ISR aircraft, and SOF aircraft in CODE analysis. Only bomber and fighter sorties are included in most of the notional results in this report, since they are the only non-tanker aircraft in the notional scenario.

---

[37] As noted in Chapter Four, the user can choose which types of sorties to include in the performance metric.

Finally, TAB-ROM's default objective function maximizes sortie generation across a range of user-defined attack vectors. As shown previously in Figure 4.14, an adversary's sortie degradation capability can vary significantly between different targeting plans. In that figure, Red's best apparent strategy is to target all cruise missiles at fuel storage and all ballistic missiles at runways. At first glance, this might suggest to Blue that the best investment strategy is to spend every available dollar on additional storage tanks and runway repair. However, the adversary does get a vote in his own shot plan. If Red were to shift tactics and heavily target parked aircraft, Blue would find he has ample fuel and runways available, but fewer aircraft to send to the fight. The consequences from over-predicting an adversary's attack plan are clearly problematic.

Given that no one can know with certainty an enemy's strategy until it eventuates, it may be prudent to invest in a mitigation portfolio that is as resilient as possible against a range of attacks against runways *and* parking *and* fuel systems. Diversifying a portfolio will help to protect the base as a whole and mitigate risk of damage to its resources. While such an approach may not be the best against any *individual* attack vector, the portfolio can be designed to generate the highest number of average sorties when played against a *range* of attack vectors in the aggregate.[38]

To this end, TAB-ROM tests many candidate investment portfolios against multiple points on the sortie generation surface (such as the one depicted in Figure 4.14). TAB-ROM evaluates the performance of a portfolio against an enemy levying all his ballistic weapons against runways, then all ballistics against parked aircraft, then all cruises against fuel, all cruises against parking, and at many points between these extremes. Note that TAB-ROM's default setting considers each of these attack vectors to be equally weighted. After assessing a portfolio against *each* of these attacks, TAB-ROM computes the average sortie generation performance against *all* of them. Designs using this approach then serve as a hedge against uncertainty in an adversary's attack plan. Such an assessment protocol is known in the optimization literature as *robust*.

The user has the option of changing this equal weighting of all attack vectors within TAB-ROM. For example, an analyst may conclude that the allocation of all cruise missiles against fuel storage runs counter to Red firing doctrine, and would thus be a very unlikely attack vector. The user can opt to reduce the weight of such attack vectors—or even remove them entirely—in the objective function. Similarly, the user can increase the weighting of specific attack vectors to represent more-likely strategies. Yet another approach is to limit the objective function to that portion of the tradespace that is within a certain percentage of the minimum. This would focus the optimization on a range of Red's best attack vectors.

So far, this discussion has centered on TAB-ROM's default objective function. The user has the option, however, of adjusting each of the weights discussed above. As touched on, the user

---

[38] A portfolio approach has advantages beyond sortie generation. For example, investing in a broad range of resiliency options can signal that Blue is committed to supporting the theater, thus helping to deter adversary aggression.

can opt to change the default weights of (1) sorties of aircraft types, (2) sorties launched from different bases, and (3) sortie outcomes from different adversary attack vectors. For example, the user might wish to shift the focus of the optimization away from C2ISR platforms and toward strike aircraft, or to place more importance on close-in main operating bases than far-away transportation hubs, or perhaps even to downplay scenarios where Red targets cruise missiles against parked aircraft. Each of these options affords the user significant opportunities to run sensitivity analyses to test scenario assumptions, portfolio outcomes, and alternative performance metrics (e.g., damaged aircraft).

Framing the Optimization

Searching the space of investment possibilities for the best portfolio is daunting. This section addresses such optimization problems and their relevance to base resiliency by discussing the advantages and shortcomings of four key methodologies: enumeration, mixed integer programming, mixed integer nonlinear programming, and heuristic programming. We chose not to use the first three, for reasons given below, and selected a heuristic programming approach— the *genetic algorithm*—for TAB-ROM.

Enumeration

Perhaps the simplest approach for searching for an investment portfolio with the highest sortie generation is enumeration. In this technique, the user simply lists all investment possibilities, assesses whether they satisfy his budget criterion and constraint set, and computes the objective function. This method is easy to envision and the simplest to set up in, for example, a spreadsheet environment. Another key benefit of enumeration is the user's ability to readily sift through all cases to determine which investment portfolio offers the best sortie generation.

Enumeration analysis is readily run for a small number of bases with a limited number of investment opportunities at each base. For example, consider a beddown where the user would like to evaluate the utility of investing in ADR teams at each base. As shown in Table 5.1, there are five levels of capability considered in the illustrative runs (one of conventional CE, four using Advanced ADR technology). To assess the robust performance of each investment, the user chooses to consider 25 evenly distributed points on the adversary attack surface (depicted on the x and y axes of Figure 4.14). TAB-VAM typically requires roughly 5 seconds to run each of these points using a 3.0 GHz CPU. The overall requirement for run time depends on the number of bases in the beddown, and Table 5.2 depicts the relationship between base count and run time.

Table 5.2. Run Times for Enumeration of Portfolio Options

Base Count	Solution Time (hrs)
1	0.2
2	0.9
10	339,084
20	3,311,369,154,188

The computation time required to evaluate options in the enumeration approach grows quite rapidly with an increasing number of bases under consideration. While it may be the simplest method to set up, enumeration will not prove to be an efficient analytic approach when the user needs to evaluate sortie generation at even a modest number of bases.

Mixed Integer Programming

The next simplest optimization approach is known as mixed integer programming (MIP), basic elements of which were discussed in Chapter Three in the discussion on the ROBOT model. MIP is an offshoot of linear programming (LP), which was developed in the 1940s by George Danzig to solve problems relevant to military logistics. Setting up MIP is readily achieved in a programming language such as GAMS, and commercial solver packages, such as Cplex, can typically complete MIP in minutes or hours. When enumeration proves cumbersome, users often try to set up their problems in the MIP framework. Much like enumeration, MIP possesses one key analytic advantage: The solution to MIP is guaranteed to be the optimal one.

A few key conditions need to be satisfied in order for a problem to qualify for MIP: (1) All the variables in the problem must either be continuous (e.g., portfolio cost) or be integers (e.g., the number of shelters or fuel tanks to be built at a base), (2) all the problem's constraints must be expressed as linear algebraic functions (e.g., establishing a ceiling on the number of FORCE kits bought), and (3) the objective function must be a linear algebraic function (e.g., the sum of all operational sorties at all bases, averaged across all adversary attack options).

At first glance, it might appear that the TAB-ROM framework satisfies the MIP criteria. Unfortunately, TAB-ROM's structure violates the third condition, that the objective function is algebraically linear. While the average of sorties across attack options is, in and of itself, linear, the computations required to determine the sortie count against each adversary attack vector are not. Recall that TAB-ROM uses TAB-VAM to calculate the sortie generation that results from each investment portfolio TAB-ROM wishes to evaluate. TAB-VAM is a complex amalgam of conditional logic (e.g., if a cruise missile defender failed to interdict its target, then the missile may prosecute its intended target) and random processes (e.g., pick a random number to determine whether that cruise missile's CEP was sufficient to strike its target). TAB-VAM, like most complex Monte Carlo simulation models, is definitively nonlinear.

Mixed Integer Nonlinear Programming

With the insight that the TAB-ROM/TAB-VAM link for computing the objective function is nonlinear, it seems reasonable that TAB-ROM could be solved using mixed integer nonlinear programming (MINLP). MINLP is related to MIP, only without MIP's requirements for linear behavior in the constraints and objective function. Problems of this nature are, as with MIP, readily set up in a language such as GAMS, and many commercial solvers, such as DICOPT, SBB, and LINDO, can solve MINLP algorithms.

MINLP, however, suffers from two key shortfalls in the context of our problem. First, it typically requires an extremely large (but indeterminate) number of calculations of the objective function. Recall that TAB-ROM's objective function of total sortie count resides within TAB-VAM, which, as noted earlier, requires roughly 5 seconds on a modern CPU to calculate the theater-wide sorties from a single adversary attack vector. To secure a robust evaluation for an investment portfolio, TAB-ROM typically assesses the portfolio effect against 25 or more adversary attack vectors. Consequently, each evaluation of the objective function requires 1.25 minutes. Even if the MINLP algorithm required only 10,000 evaluations of the objective function, the total solution time would still exceed eight days.

The MINLP framework lacks another major advantage of enumeration and MIP: The MINLP solution offers no mathematical guarantee that it is truly the best option available. It can be a very good answer, perhaps even the best, but the analyst cannot be certain of the solution quality. The challenge stems from MINLP's inability to efficiently search what is known in the optimization lexicon as a *nonconvex* space. Nonconvexity refers to any "bumpy" or unsmooth surface, like a sine wave, a mountain range, or the surface depicted in Figure 4.14. On that plot, there are several ridges and valleys contributing to the surface's nonsmooth character.

Consider how a typical MINLP maximization algorithm would approach finding the tallest peak in a mountain range. The solver begins by generally randomly selecting a single point in the range, which only has visibility to nearby points. Consequently, the optimization proceeds by climbing the nearest and steepest slope. Once it reaches the summit of that slope, it concludes the search and returns the coordinates of that peak to the user. The problem in a nonconvex space is that there may be a taller mountain somewhere else in the range, but the user would not know it without restarting the algorithm from different starting points to get a feel for the overall terrain. (Moreover, the peak may not even represent a feasible solution, and without restarting from a new location, the solver may be unable to return a meaningful solution at all.) The technical lexicon refers to this phenomenon as being trapped in a *local optimum*, whereas the ultimate goal of MINLP is to uncover the highest peak in the space, known as the *global optimum*.

Heuristic Programming

The paired problems of slow searches and becoming trapped at local optima—let alone the issue of finding solutions at all—have been recognized by the optimization literature for decades and have received extensive treatment. General solution methodologies involve using specialized, faster, and approximate search techniques, or *heuristics*, to trade off for the analytic precision of mathematical programming methods like MIP and MINLP. Heuristics are complex, rule-based searches that typically mimic natural phenomena. Popular techniques include neural networks, based on precepts from the field of neuroscience, and simulated annealing, which mimics the process of crystal formation in metallurgical annealing. TAB-ROM uses yet a third popular approach, the genetic algorithm, addressed in the next section.

The Genetic Algorithm

To overcome the limitations of linear and nonlinear programming approaches, TAB-ROM uses a genetic algorithm (GA) as its search engine. A GA mimics evolutionary processes to span the solution space in search of the best solution available. In the context of the CODE analysis, the GA begins with a random selection of candidate investment portfolios, known as a *population*. TAB-ROM passes each of these investment strategies, or *chromosomes*, from the population to TAB-VAM to assess how well the strategy performs in terms of the sorties it can generate. This metric is known as the *fitness value* of that chromosome. Should an investment strategy violate the budget constraint (or any other bounds the user should choose to impose), TAB-ROM levies a penalty against its fitness score. Such a penalty degrades the chromosome's overall fitness, but any useful traits in the portfolio the chromosome represents can still be preserved within the population.

The GA then combines (*mates*) random features from the best-performing chromosomes in the population to create a new *generation* of candidate investment strategies. For example, one well-performing candidate may have opted to put an Advanced ADR team at Air Base X and a hardened fighter shelter at Air Base Y. The offspring of these chromosomes may contain both mitigations. Occasionally, the GA will insert into the population a chromosome that *almost* copies another successful member of the group. The deviation will include a combination of a random investment at a random base (a *mutation*) that will slightly alter the original chromosome's overall investment strategy. Such mutations introduce a small shift in the solution, which helps to prevent the GA from focusing solely on a local optimum. The GA then continues this process of combining investment strategies to evolve new, improved ones until a user-defined cutoff, generally determined by the passage of a given number of generations without any significant improvement in the fitness score. TAB-ROM then outputs the base-by-base investment strategy, which the user can subsequently input into TAB-VAM to determine its base-specific effects on sortie generation.

The GA underpinning TAB-ROM is currently written in FORTRAN and employs Perl scripts to read outputs from TAB-VAM and to render the GA's own outputs readable by TAB-VAM. TAB-ROM is typically set up with population sizes of ten to 20 portfolios and runs for about 100 generations. Depending on the size of the beddown being studied, TAB-ROM run times usually range between 12 and 48 hours. At the conclusion of a TAB-ROM run, the user has the option of continuing the search from the best candidate population evolved. In so doing, the user can determine whether TAB-ROM had converged sufficiently to an appropriate investment portfolio.

Illustrative Model Run

For the rest of this chapter, we will examine TAB-ROM outputs that draw on the notional scenario described in the appendix. First, we will discuss the basis for a set of constraints that apply to the scenario. We will then explore the cost/effectiveness trade-offs between passive defense procurements in ADR (such as Advanced ADR teams), notional aircraft shelters, and fuel infrastructure enhancements (such as FORCE kits and static storage tanks). We will show how different investment mixes—and their associated sortie generation—vary across different budget levels. Given that Blue operational plans may call for allied participation as part of a dispersed basing posture (as discussed in Chapter Two), we will also show the effects of allied contributions. Finally, we will illustrate how TAB-ROM results can help identify the point at which further increases in budget will yield diminishing returns. As in previous chapters, the aim of this section is to highlight TAB-ROM's capabilities rather than to draw concrete policy recommendations from this highly notional scenario.

Scenario Constraints

If tensions are expected to rise in the region, a prepared Blue force would conduct many of the analyses discussed in previous chapters. Namely, Blue would evaluate various beddown plans, the UTC requirements for those plans, FSLs for storing the WRM needed to execute them, and the cooperative security agreements that would be useful in supporting Blue forces. As discussed previously, each of these analyses would help to better inform Blue on the materiel requirements and political partnerships needed to successfully prosecute a conflict, should the need eventuate. Following these analyses with the simple act of investing in materiel and infrastructure, at both FSLs and potential operating locations, may even successfully signal Blue's intent to protect the region and demonstrate its commitment to allies. Let us assume that Blue has conducted the analyses illustrated in previous chapters and is prepared to use the outcomes to develop an investment plan for the theater.

Chapter Four highlights several analyses important to the notional scenario. In particular, Figure 4.15 shows that distributing combat assets to the outer islands, Bases 5–7, and incorporating allied forces centered at Base 3 play important roles in complicating Red's targeting strategies. The additional targets diminish the impact of Red's quiver at any one base. Moreover, when Blue is able to draw on allied support in the region, those partners can bring combat power to the fight—and possibly additional capital for investment in mitigation resources. Consequently, the baseline for force posture in TAB-ROM analyses discussed here will be the Dispersed Beddown.

As shown in Chapter Two, the UTC requirements for the Dispersed Beddown are only modestly greater than those for the Consolidated Beddown. Let us assume that Blue's Air Force has sufficient capability to support those requirements. If shortfalls do arise, say, in security forces, fuels management, or civil engineering, let us also assume that the Joint community will lend its support to fill any gaps. The Army and Navy, for example, could draw on their Corps of Engineers and Seabees to complement the Air Force's pool of CE assets. If shortfalls are quickly recognized, the Air Force could train the other forces in skills specific to its own needs, such as the task of EOD in a runway environment or perimeter protection at airfields. Because of the relatively small scale of the notional scenario's conflict and access to the Joint pool of manpower and materiel, **we will assume that this fight would not be UTC limited**.

As discussed in Chapter Three, maneuvering from the Dispersed Beddown will necessitate higher costs than the Consolidated Beddown for transporting WRM into all operating locations. We will assume that Blue has secured sufficient funding for dispersed operations and has sized its FSLs and WRM appropriately for a conflict of this scope. Furthermore, we assume that Blue has secured appropriate transportation options, such as airlift and rapid surface vessels, to close its TPFDD prior to initiation of the conflict.

Given the island-based geography of the scenario, let us assume that Blue's operating locations are space-constrained with respect to new infrastructure. In particular, we will assume that each base can receive no more than 10 million gallons of above-ground fuel storage. This will limit procurement to no more than eight 1.2-million-gallon static fuel tanks at each location. Space is also restricted for placement of FORCE kits. The area at each base allotted for setting out FORCE will be limited to the bladders and pumps sufficient for three days of combat flight operations. Furthermore, these space restrictions will limit the construction of aircraft shelters to two-thirds of the aircraft bedded down at each location. These ceilings are enumerated in Table 5.3.

Table 5.3. Space Constraints on Fuel Storage and Shelter Procurement

Base	Maximum 1.2 Mgal FORCE Kits	Max 1.2 Mgal Fuel Storage Tanks	Maximum Shelter Count
1	1	8	24 (small)
2	1	8	12 (small)
3	1	8	8 (small)
4	1	8	12 (small)
5	4	8	22 (large)
6	1	8	6 (large)
7	1	8	6 (large)

Finally, we will assume that procurement of and upgrades to Advanced ADR teams are possible at each base. Note that four operating locations (Bases 1, 5, 6, and 7) each already possess small Advanced ADR teams. TAB-ROM will be allowed to augment the capability at these bases with additional hardware to form larger teams. However, given that the materiel for a small team is already on-hand, larger teams can be procured at a cost discount equal to the cost of the small team. For example, if TAB-ROM sees advantage in upsizing a small team to a very large capability, from Table 5.1, the net cost will be $(60 – 15) million = $45 million. If TAB-ROM wishes to procure an Advanced ADR team to replace existing CE capability found at Base 2, 3, or 4, it will pay the full price of the Advanced ADR team as shown in that same table.

Optimal Investment Mix at $1 Billion

First, let us compare two sortie generation performance surfaces for the Dispersed Beddown: one with no additional investment (Figure 5.1) and another with a one-time, TAB-ROM-optimized investment of $1 billion (Figure 5.2).[39] We assume that Red's targeting strategy is to attack all air bases in proportion to the number of aircraft at each base. Investments will be allowed at all bases except for the allied location at Base 3 (we will return to the effect of allied investments shortly).

---

[39] To demonstrate the effects of investment, the vertical scale of Figures 5.1 and 5.2 is greater than that of Figure 4.14. Thus, the surface on Figure 5.1 appears to be lower and flatter than that on Figure 4.14. In fact, the surface has a higher maximum, corresponding to the second box in Figure 4.15.

Figure 5.1. Average Sortie Generation for Dispersed Beddown (No Additional Investment)

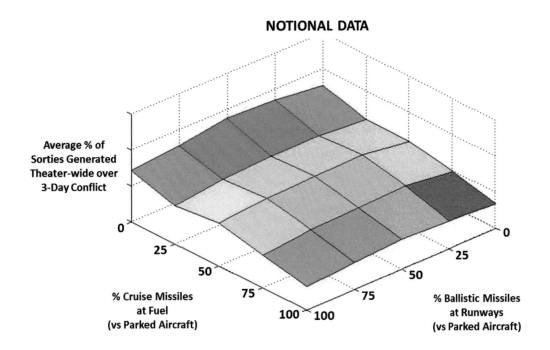

Figure 5.2. Average Sortie Generation for Dispersed Beddown ($1 Billion Investment)

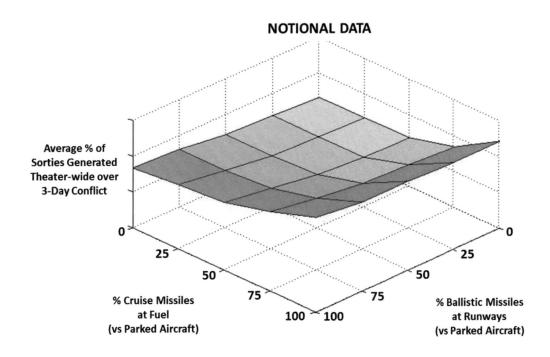

In the case with no investment, Red achieves the largest degradation in Blue sortie generation by targeting all its cruise missiles at fuel storage and all its ballistic assets at runways. With $1

billion in investment in damage avoidance and mitigation strategies (detailed in the next section), the average height of the surface rises. Mitigations lift the surface by adding damage recovery capability to runways (with Advanced ADR teams), damage sustainment and avoidance to fuel systems (by dispersing bladders and adding redundant tanks), and damage avoidance to parked aircraft (with hardened shelters).

A $1 billion investment significantly improves Blue's resiliency to attacks on fuel and runways and appreciably upgrades survivability to ballistic missile attacks on parking areas. In this notional scenario, Blue's investments in ADR and fuel systems offer the most cost-effective payoff in sortie generation, whereas hardened shelters appear to offer a somewhat lower marginal advantage in terms of sorties protected per investment dollar, as discussed in the next section.

Optimal Investment Mixes at Various Budget Levels

While a $1 billion investment offers significant benefit, as shown above, a more important questions is: What is the "right" level of investment? That is, would further investment afford even greater protection, or would additional capital provide minimal additional return? Decisionmakers and stakeholders frequently face this problem, and it can be quite challenging. TAB-ROM can help answer this question by calculating the average sortie generation (and optimal portfolios) at many investment levels. The result is a curve representing the trade-off between average sortie generation capability and investment level, as shown in Figure 5.3.

Figure 5.3. Trade-Off Between Level of Investment and Average Sortie Generation

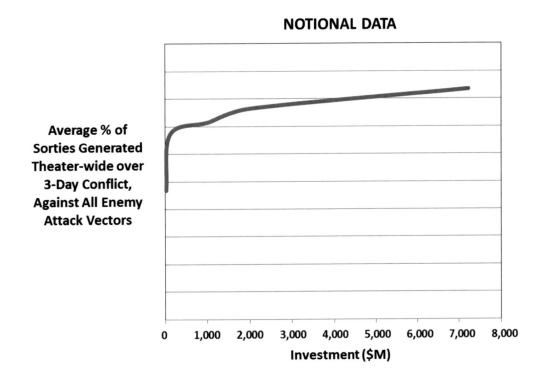

The figure shows that, for the notional scenario, a limited investment of about $100 million produces a rapid improvement in average sortie generation across all attack vectors. The effect of further investments decelerates until $1 billion is reached, jumps a bit from $1 billion to $2 billion, and increases slowly for investments above the $2 billion level. Investments cease at approximately $7.2 billion, as this is the point at which ADR capability is maximized at each base, and space restrictions limit further procurement of hardened shelters and fuel storage.

While Figure 5.3 indicates the behavior of the optimized investment package, it does not portray the relative investment levels between the three major mitigation categories of fuel, ADR, and shelters. Figure 5.4 depicts TAB-ROM's breakout between these investment classes. The figure shows that, in the notional scenario, optimal procurements of ADR and fuel storage taper off beyond a total budget of $2 billion. Investment in shelters, however, climbs steadily as budget levels are allowed to grow. However, the marginal advantage of shelters is relatively low, as indicated by the fact that the curve in Figure 5.3 has a much greater increase between $0 and $2 billion than between $2 billion and $7.2 billion.

The following sections detail investments in each resource at various budget levels.

Figure 5.4. Division of Investment Among Mitigation Categories

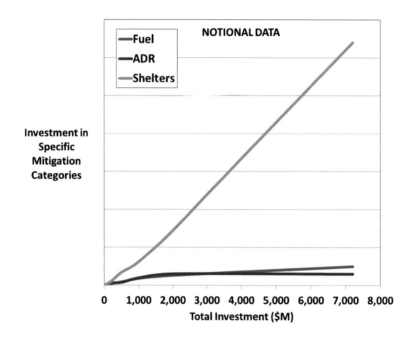

Fuel Storage and Dispersal

Relative to ADR and hardened shelters, purchasing fuel storage and dispersal offers an inexpensive strategy for improving a base's resiliency to attack. As fuel is required by every sortie on a base, an operating location can be especially vulnerable to a significant attack on its fuel storage. Consequently, investments in fuel storage and dispersal are essentially blanket insurance policies that can help to protect all of a base's sorties.

TAB-ROM's specific fuel-system buys at key investment levels for the notional scenario are presented in Table 5.4. Note that the table presents only the fuel storage components of the optimal investment in ADR, shelters, and fuel. At lower budget amounts, TAB-ROM procures FORCE kits in greater numbers than static tanks. As budget levels increase, TAB-ROM hits the procurement ceiling for FORCE kits and then proceeds to buy increasing numbers of above-ground tanks. Beyond the $2 billion investment level, the overall fuel storage investment sees marginal return in terms of sortie generation. In this investment regime, TAB-ROM sees greater benefit in upgrading ADR and procuring shelters. However, TAB-ROM will continue to buy static tanks until it hits the constraint of available space at each base.

66

Table 5.4. Additional Tanks/FORCE Kits Procured at Varying Investment Levels (Notional)

Base	$0	$100M	$1B	$2B	$7.2B
1	0/0	0/1	6/1	6/1	8/1
2	0/0	0/1	1/1	2/1	8/1
3	0/0	0/0	0/0	0/0	8/0
4	0/0	1/0	4/1	7/1	8/1
5	0/0	0/2	2/4	3/4	8/4
6	0/0	0/1	2/1	3/1	8/1
7	0/0	1/0	1/1	3/1	8/1

ADR

Just as with their sensitivity to fuel attacks, bases can be particularly vulnerable to attacks on runways. Each aircraft bedded down at an operating location requires an intact runway for sortie generation. Larger aircraft, such as bombers, tankers, and airlift, are especially susceptible to attacks on runways, as they require greater MOS for their takeoffs and landings. Thus, advanced ADR capability can provide another form of relatively inexpensive insurance that can help to protect every sortie at a base.

TAB-ROM's specific upgrades to ADR capabilities at key investment levels are shown in Table 5.5. As the budget increases, TAB-ROM gradually improves the ADR capability at each base. By the $2 billion investment level, TAB-ROM has upgraded each operating location to the Very Large Advanced ADR capability (with the exception of the allied site at Base 3, which has not yet been allowed investment opportunities). Also of note is the model's decision to rapidly upgrade the ADR capability at Base 1, which is susceptible to the highest levels of runway attack (as shown in Figure 4.8), and Bases 5–7, where all of the large aircraft in the scenario are bedded down.

Table 5.5. ADR Capability Procurement at Varying Investment Levels (Notional)

Base	$0	$100M	$1B	$2B	$7.2B
1	S	M	VL	VL	VL
2	CE	CE	S	VL	VL
3	CE	CE	CE	CE	CE
4	CE	S	S	VL	VL
5	S	S	M	VL	VL
6	S	S	M	VL	VL
7	S	S	M	VL	VL

NOTE: S/M/L/VL = Small/Medium/Large/Very Large Advanced ADR.

Hardened Aircraft Shelters

Of the three mitigation options TAB-ROM explores here, hardened aircraft shelters offer the smallest degree of sortie protection per investment dollar. However, a shelter can offer significant defense against broad-area attacks that can damage large numbers of aircraft parked on an open apron. Moreover, a shelter may prove to be a relatively inexpensive insurance policy for aircraft that can be quite expensive and of high value in the conflict, such as assets constructed with low-observable materials or designed with long-range strike capabilities.

Table 5.6 shows TAB-ROM's spend pattern on hardened shelters across the investment spectrum. At lower levels of overall investment, TAB-ROM does not detect utility in procuring significant numbers of shelters over the base-wide protection offered by upgrading fuel storage and ADR capabilities. However, by the $1 billion investment point, TAB-ROM adds a significant number of small shelters for fighters closer to the conflict.[40] Large shelters do not enter the mix until about the $2 billion level of investment. Beyond this point, TAB-ROM procures more shelters, both large and small, until the shelter capacity at each base is reached.

Table 5.6. Hardened Aircraft Shelter Procurement at Varying Investment Levels (Notional)

Base	Shelter Size	$0	$100M	$1B	$2B	$7.2B
1	Small	0	0	7	24	24
2	Small	0	2	12	12	12
3	Small	0	0	0	0	0
4	Small	0	0	12	12	12
5	Large	0	0	0	3	22
6	Large	0	0	0	0	6
7	Large	0	0	0	0	6

The table highlights another interesting feature underpinning the procurement of small shelters for fighter aircraft. Note that the first shelters TAB-ROM buys are at Base 2, and the maximum number is selected for that location by the $1 billion investment mark. There is a form of positive feedback loop in protecting the cruise missile defenders located here. By safeguarding these aircraft against parking attacks, these fighters are more likely to survive the entirety of the three-day conflict. Their survival ensures the ability to repeatedly engage more of Red's cruise missiles. That continuing engagement thus offers ongoing protection for the cluster of Bases 2–4 against this threat. The defenders thereby help to ensure the sortie generation of all aircraft in the cluster. Consequently, shelters bought in sufficient quantity can afford a blanket insurance-like

---

[40] At the $1 billion investment level, hardened shelter procurement hits its maximum at Bases 2 and 4. From the appendix, note that each base has 18 aircraft, one-third of which are airborne at any given time. Thus, TAB-ROM sees no advantage in procuring shelters to protect more than two-thirds (i.e., 12) of the fighters. At the $7.2 billion investment level, TAB-ROM invests in the maximum number of shelters at each operating location.

policy similar to that offered by ADR and fuel mitigations. While the base-wide insurance obtainable by shelters is more expensive than that provided by ADR and fuel, the investment can protect multiple bases in a regional cluster.

Allied Investments

To this point, we have not discussed the sortie generation protection that allies can provide by their own investments in mitigations. Under a cooperative security agreement, such as NATO's Partnership for Peace, allies can share defense techniques and practices, and then exercise them together to ensure successful integration. In the notional scenario discussed here, Blue could share the technologies behind Advanced ADR, FORCE, and advanced hardening capabilities with the ally at Base 3 in return for guarantees of base access, overfly rights, and combat support during the conflict. That ally could then allocate from its own defense budget to procure these capabilities at its operating location. This could be seen as a prudent defense policy: If an adversary were willing to attack Blue's assets on the ally's sovereign territory (e.g., Bases 2 and 4), it would not be surprising to imagine that Red would be willing to attack that ally's military assets at Base 3 as well.

Similar to the results shown in Figure 5.3, Figure 5.5 depicts the outcome if TAB-ROM were allowed to manage two separate budgets – one for the allied investments at Base 3, another for Blue at all other bases. We assume that investments at Base 3 are made with allied funds. For these TAB-ROM runs, the ally was allowed to invest no more than Blue. The ally was also not allowed to violate any of the scenario's procurement constraints, which place restrictions on the number of hardened shelters, FORCE kits, and above-ground fuel tanks. With these cost constraints, the ally is able to match Blue investments up until approximately $300 million, at which point the ally will have maximized its purchasing potential at Base 3.

Figure 5.5. Comparison of the Effect of Blue-Only Investments with Blue and Allied Investment

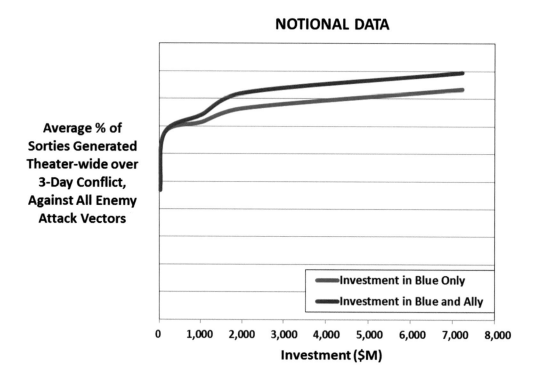

NOTIONAL DATA

Average % of Sorties Generated Theater-wide over 3-Day Conflict, Against All Enemy Attack Vectors

Investment in Blue Only
Investment in Blue and Ally

0   1,000   2,000   3,000   4,000   5,000   6,000   7,000   8,000

Investment ($M)

The figure shows that the characteristic shape of the investment response curve has not changed from the Blue-only procurement case. However, by allowing the ally to purchase damage mitigation and recovery technologies, the ally is able to increase the theater-wide average sortie generation potential by several percentage points. This increase is achieved with allied investments as low as $200 million. Given that the ally is capable of generating 11 percent of the scenario's overall operational sorties,[41] the increased performance represents a significant capability boost for the ally.

Table 5.7 shows TAB-ROM's spend pattern for allied investment. The Blue-only investment does not change from that shown in Tables 5.4–5.6.

---

[41] This is based on Table A.3.

Table 5.7. Allied Procurement Plan at Varying Investment Levels (Notional)

Resource	$0	$100M	$250M	$300M
New fuel tanks	0	2	6	8
FORCE kits	0	1	1	1
ADR	CE	M	VL	VL
Small hardened shelters	0	2	6	8

The table shows that, at relatively modest levels of investment, TAB-ROM upgrades the ally's ADR capability and procures a FORCE kit to distribute fuel on-base. The number of hardened aircraft shelters and additional fuel storage tanks entering the investment portfolio increases gradually. These resources eventually hit their ceiling constraint at the maximum investment level.

Conclusion

In concert with TAB-VAM, TAB-ROM identifies robust investment portfolios that maximize theater-wide sortie generation by mitigating risks from a spectrum of adversary attack vectors. It details what resources to procure and where to place them. TAB-ROM can help stakeholders visualize the cost-effectiveness trade-offs at varying levels of investment and the point at which further investment yields diminishing returns. Finally, it shows how allies can support Blue capability with their own capital contributions. As with TAB-VAM, TAB-ROM can be expanded to examine a wider range of resources and investment options.

6. Conclusion

This research report describes the suite of models that PAF is using to analyze support for combat operations in denied environments. It has described the major inputs, algorithms, and outputs of each model and illustrated its use in a notional scenario.

In the actual CODE analysis, we use these models to examine a variety of questions of interest to force planners. By providing insights into combat support requirements, vulnerabilities, resiliency, and capability trade-offs, the modeling framework is helping to inform the development of operational and support CONOPS in denied environments, current and future investment decisions, area of responsibility basing strategies, discussion within the Secretary of Defense's Management Action Group, and Air Force inputs to the Quadrennial Defense Review.

We are in the process of further developing areas of modeling capability to extend the breadth and fidelity of PAF's analytic framework. These areas include

- improving the modeling fidelity of active missile defense assets, including expanded interceptor inventory and the vulnerability of launch platforms to adversary attack
- expanding the models' visibility into Blue munitions, including factors such as theater inventory, altering munitions storage CONOPS, and assessing the vulnerability of munitions storage areas
- broadening TAB-VAM's assessments of fuel infrastructure to include attacks on the broader supply chain of intra-theater fuel transportation, as well as receipt and distribution at individual operating locations
- extending the assessment of attacks to include the broader fuel supply chain, to include fuel receipt and distribution
- assessing the impacts of adversary attacks on maintenance personnel and repair facilities on sortie generation capability
- including the role of cyber attacks on the disruption of combat support operations, such as the delay and diversion of WRM prior to and during a conflict.

Many of these additional modeling features will draw on expertise from the Joint community, given that factors such as fuel, electricity, materiel delivery, and munitions influence more than just the Air Force.

Appendix. Notional Scenario

To demonstrate the capabilities of the PAF models and highlight some of the investment trade-offs that can be explored with this framework, we developed a notional scenario for the examples shown in this report. The scenario is generic, with no actual countries, air bases, or weapons being used as inputs. The use of a generic, custom scenario made it possible to efficiently capture many aspects of the modeling tradespace, such as dispersal of bases, the use of allied bases, and varying damage mitigation resources. The scenario is for a three-day, short-term conflict.

This appendix describes the major scenario assumptions, which consist of the adversary missile quiver, Blue beddowns, and Blue base infrastructure and damage repair capabilities.

Adversary Missile Quiver

The adversary in this scenario has short-, medium-, and intermediate-range ballistic missiles as well as both ground- and air-launched cruise missiles. Although TAB-VAM has the capability to model several different launch sites, we used a single site in this scenario for simplicity. The inventory and ranges of the adversary missiles are given in Table A.1.

Table A.1. Adversary Missile Ranges and Inventory

Missile	Max Range (km)	Total Inventory	Allocated to Bases (60%)
SRBM #1	1000	100	60
MRBM #1	3000	50	30
IRBM #1	5000	10	6
GLCM #1	2500	150	90
ALCM #1	5000	150	90

Note that a "strategic withhold" is applied to the inventory values in Table A.1, since the adversary would likely not use up its entire arsenal on a single conflict. We assume a 40 percent strategic withhold in this scenario, which leaves 60 percent of the inventory to be used against air bases. From the missiles in TAB-VAM, half are launched on the first day, leaving 25 percent for each of the remaining two days. We assume that each Red missile has a 100 percent reliability rate (i.e., there is zero chance of the missile failing from launch to arrival). We also assume that Red has prior knowledge of the aircraft beddown; therefore, we specify that TAB-VAM will allocate missiles based on the total number of aircraft at each base (as described in Chapter Four).

Beddown

The scenario offers two beddown alternatives: a "Consolidated Beddown" consisting of five bases and a "Dispersed Beddown" consisting of seven bases.

Figure A.1 shows the Consolidated Beddown. The theater comprises the adversary mainland and a succession of islands at distances of 800, 1,600, and 3,200 km. In the Consolidated Beddown, Blue and allied aircraft are positioned at five bases: one within SRBM range, three more within MRBM range, and one more within IRBM range. Base 3 is an allied base. The number and types of aircraft at each base are listed in Table A.2. In this and the Dispersed Beddown, we assume that one-third of aircraft are in the air at any given time.

Figure A.1. Consolidated Beddown

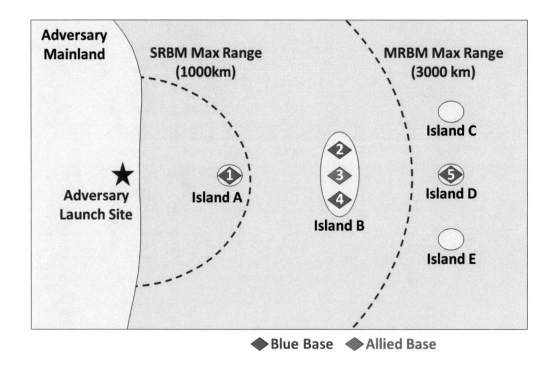

Table A.2. Consolidated Beddown

Base	Blue or Allied	Distance from Mainland (km)	Fighters		Bombers		Tankers	
			Number	Sortie Rate	Number	Sortie Rate	Number	Sortie Rate
Base #1	Blue	800	36	2	-	-	-	-
Base #2	Blue	1600	18	2	-	-	8	1
Base #3	Allied	1600	12	2	-	-	-	-
Base #4	Blue	1600	18	2	-	-	8	1
Base #5	Blue	3200	-	-	24	2	8	1

Figure A.2 shows the Dispersed Beddown. In this case, the Blue tankers are moved from Bases 2 and 4 on Island B to two new bases (6 and 7) in the outer island chain, outside the MRBM threat ring. Base 3 remains an allied base. The number and types of aircraft are listed in Table A.3.

Figure A.2. Dispersed Beddown

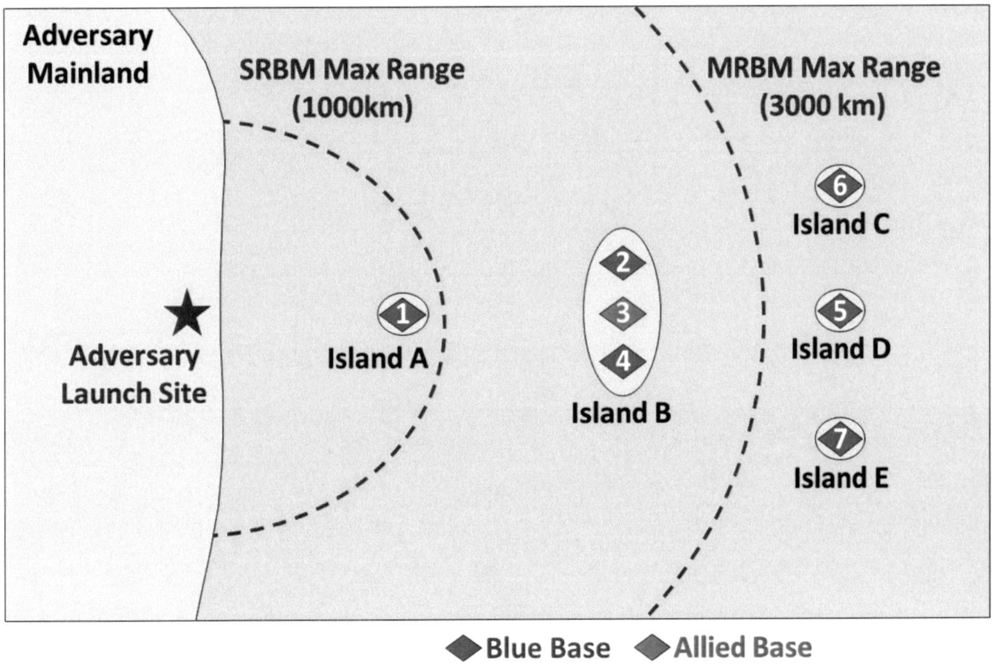

Table A.3. Dispersed Beddown

Base	Blue or Allied	Distance from Mainland (km)	Fighters		Bombers		Tankers	
			Number	Sortie Rate	Number	Sortie Rate	Number	Sortie Rate
Base #1	Blue	800	36	2	-	-	-	-
Base #2	Blue	1600	18	2	-	-	-	-
Base #3	Allied	1600	12	2	-	-	-	-
Base #4	Blue	1600	18	2	-	-	-	-
Base #5	Blue	3200	-	-	24	1	8	1
Base #6	Blue	3200	-	-	-	-	8	1
Base #7	Blue	3200	-	-	-	-	8	1

Base Infrastructure and Damage Repair Capabilities

We make a range of baseline assumptions regarding the infrastructure, missile defense capability, and damage repair capabilities at each base. We assume that one Blue fighter squadron at Base 2 is used in a missile defense role. Each fighter carries ten interceptors, each assumed to have a 50 percent single-shot probability of kill. Also, the missile defense fighters are assumed to have a 50 percent probability of detection of incoming missiles. We treat Bases 2, 3, and 4 as a single missile defense cluster; therefore, the squadron at Base 2 will engage enemy cruise missiles fired at any of these three bases, giving equal priority to each enemy missile fired at that cluster.

Tables A.4 and A.5 show the other assets at the bases, which primarily consist of fuel and ADR teams. We assume that no shelters are included in the baseline case.

Table A.4. Consolidated Beddown Infrastructure and Damage Repair Capabilities

Base	Blue or Allied	Airfield Damage Repair	Fuel (millions of gallons)	Runways (length x width in feet)	Parking Sections
Base #1	Blue	Small ADR	10	12,000 x 300, 12,000 x 200	19
Base #2	Blue	Conventional CE	6	11,000 x 200	65
Base #3	Allied	Conventional CE	6	11,000 x 200	65
Base #4	Blue	Conventional CE	6	11,000 x 200	65
Base #5	Blue	Small ADR	10	(2) 11,000 x 200	18

Table A.5. Dispersed Beddown Infrastructure and Damage Repair Capabilities

Base	Blue or Allied	Airfield Damage Repair	Fuel (millions of gallons)	Runways	Parking Sections
Base #1	Blue	Small ADR	10	12,000 x 300, 12,000 x 200	19
Base #2	Blue	Conventional CE	6	11,000 x 200	65
Base #3	Allied	Conventional CE	6	11,000 x 200	65
Base #4	Blue	Conventional CE	6	11,000 x 200	65
Base #5	Blue	Small ADR	10	(2) 11,000 x 200	18
Base #6	Blue	Small ADR	0.8	9,000 x 150	8
Base #7	Blue	Small ADR	0.8	9,000 x 150	8

References

Air Force Civil Engineering Center (AFCEC), E-mail, December 3, 2012.

Amouzegar, Mahyar A., Ronald G. McGarvey, Robert S. Tripp, Louis Luangkesorn, Thomas Lang, and Charles Robert Roll, Jr., *Evaluation of Options for Overseas Combat Support Basing*, Santa Monica, Calif.: RAND Corporation, MG-421-AF, 2006. As of March 2014: http://www.rand.org/pubs/monographs/MG421.html

Brooke, Anthony, David Kendrick, Alexander Meeraus, and Ramesh Raman, *General Algebraic Modeling System: A User's Guide*, Washington, D.C.: GAMS Development Corporation, 2003.

Emerson, Donald E., *An Introduction to the TSAR Simulation Program: Model Features and Logic*, Santa Monica, Calif.: RAND Corporation, R-2584-AF, 1992. As of March 2014: http://www.rand.org/pubs/reports/R2584.html

Kolitz, S. E., "Analysis of a Maximum Marginal Return Assignment Algorithm," *Proceedings of the 27th IEEE Conference on Decision and Control,* Vol. 3, 7–9 December 1988, pp. 2431, 2436.

Lostumbo, Michael J., Michael J. NcNerney, Eric Peltz, Derek Eaton, David R. Frelinger, Victoria A. Greenfield, John Halliday, Patrick Mills, Bruce R. Nardulli, Stacie L. Pettyjohn, Jerry M. Sollinger, and Stephen M. Worman, *Overseas Basing of U.S. Military Forces: An Assessment of Relative Costs and Strategic Benefits*, Santa Monica, Calif.: RAND Corporation, RR-201-OSD, 2013. As of March 2014: http://www.rand.org/pubs/research_reports/RR201.html

Lynch, Kristin F., John G. Drew, Robert S. Tripp, and Charles Robert Roll, *Supporting Air and Space Expeditionary Forces: Lessons from Operation Iraqi Freedom*, Santa Monica, Calif.: RAND Corporation, MG-193-AF, 2005. As of March 2014: http://www.rand.org/pubs/monographs/MG193.html

McGarvey, Ronald G., Robert S. Tripp, Rachel Rue, Thomas Lang, Jerry M. Sollinger, Whitney A. Conner, and Louis Luangkesorn, *Global Combat Support Basing: Robust Prepositioning Strategies for Air Force War Reserve Materiel*, Santa Monica, Calif.: RAND Corporation, MG-902-AF, 2010. As of March 2014: http://www.rand.org/pubs/monographs/MG902.html

Morgan, Forrest E., *Crisis Stability and Long-Range Strike: A Comparative Analysis of Fighters, Bombers, and Missiles*, Santa Monica, Calif.: RAND Corporation, MG-1258-AF, 2012. As of March 2014:
http://www.rand.org/pubs/monographs/MG1258.html

Snyder, Don, and Patrick Mills, *Supporting Air and Space Expeditionary Forces: A Methodology for Determining Air Force Deployment Requirements*, Santa Monica, Calif.: RAND Corporation, MG-176-AF, 2004. As of March 2014:
http://www.rand.org/pubs/monographs/MG176.html

Stucker, James P., Ruth T. Berg, Andre A. Gerner, Amada Giarla, William L. Spencer, Lory Arghavan, and Roy Gates, *Understanding Airfield Capacity for Airlift Operations*, Santa Monica, Calif.: RAND Corporation, MR-700-AF/OSD, 1998. As of March 2014:
http://www.rand.org/pubs/monograph_reports/MR700.html

Stucker, James P., and Laura M. Williams, *Analyzing the Effects of Airfield Resources on Airlift Capacity*, Santa Monica, Calif.: RAND Corporation, DB-230-OSD, 1999. As of March 2014:
http://www.rand.org/pubs/documented_briefings/DB230.html